Nanowires—Synthesis, Properties, Assembly and Applications

MATERIALS RESEARCH SOCIETY
SYMPOSIUM PROCEEDINGS VOLUME 1144

Nanowires—Synthesis, Properties, Assembly and Applications

Symposium held December 1–5, 2008, Boston, Massachusetts, U.S.A.

EDITORS:

Yi Cui
Stanford University
Stanford, California, U.S.A.

Lincoln Lauhon
Northwestern University
Evanston, Illinois, U.S.A.

A. Alec Talin
Sandia National Laboratories
Livermore, California, U.S.A.

E.P.A.M. Bakkers
Philips Research Laboratories
Eindhoven, Netherlands

Materials Research Society
Warrendale, Pennsylvania

CAMBRIDGE
UNIVERSITY PRESS

University Printing House, Cambridge CB2 8BS, United Kingdom

One Liberty Plaza, 20th Floor, New York, NY 10006, USA

477 Williamstown Road, Port Melbourne, VIC 3207, Australia

4843/24, 2nd Floor, Ansari Road, Daryaganj, Delhi - 110002, India

79 Anson Road, #06-04/06, Singapore 079906

Cambridge University Press is part of the University of Cambridge.

It furthers the University's mission by disseminating knowledge in the pursuit of education, learning and research at the highest international levels of excellence.

www.cambridge.org
Information on this title: www.cambridge.org/9781605111162

First published 2009
First paperback edition 2012

Single article reprints from this publication are available through University Microfilms Inc., 300 North Zeeb Road, Ann Arbor, MI 48106

CODEN: MRSPDH

A catalogue record for this publication is available from the British Library

ISBN 978-1-605-11116-2 Hardback

This work was supported in part by the Office of Naval Research under Grant Number N00014-09-1-0118. The United States Government has a royalty-free license throughout the world in all copyrightable material contained herein.

CONTENTS

PREFACE

One-dimensional nanowires support the transport of charge carriers, photons and ions along their length while maintaining nanoscale effects across their diameter. The unique nanowires from versatile materials have shown great promise in nanoscale electronics, photonics, thermoelectrics, biotechnology, and energy conversion. Symposium LL, "Nanowires—Synthesis, Properties, Assembly and Applications," held December 1–5 at the 2008 MRS Fall Meeting, in Boston, Massachusetts, has provided the opportunity for discussing the critical issues related to nanowires and recent progresses in synthesis, structure, properties and devices. Specific topics of the symposium covered:

(1) Synthesis with control over composition, size, shape, position, geometry, doping, alloying and heterostructures. Materials include Group IV, III-V, II-VI, metal and metal oxide and chalcogenide materials.

(2) Properties: mechanical, electronic, optical, thermal, magnetic, ionic, phase transformational, chemical properties, etc.

(3) Assembly and integration: methods for organizing nanowires, multiple length scale pattern formation, heterogeneous integration, assembly architecture, etc.

(4) Applications: functional devices and systems for electronics, photonics, sensors, renewable energy.

The symposium consisted of 98 oral presentations, of which 14 were invited, and 142 poster presentations. The presentations were grouped into topical sessions, which covered growth mechanisms, doping, memory and logic applications, emerging applications, optical and magnetic properties, electromechanical properties, electrical and thermal transport, sensing, heterostructure synthesis, photodetection, and the synthesis and properties of metallic nanowires.

By all accounts, our symposium has been an astounding success. We received a total of 272 abstracts, second only to Symposium JJ, which also included the subject of nanowires. Due to the large number of submissions, Symposium LL included oral sessions throughout the entire 5 days of the conference, as well as 3 poster nights. Session attendance was consistently high, including several presentations with standing room only.

Many outstanding talks and posters were presented at the symposium. The presentations that particularly stand out include the invited lecture by Prof. Charles Lieber, in which he traced the developments in nanowires, highlighting the various fundamental scientific and technologically significant discoveries, such as chem/bio sensing, nanoscale light sources, quantum electronics, etc. that have been enabled by semiconductor nanowires. Another paper worth mentioning was by Naoki Fukata et al., from National Institute for Materials Sciences, Tsukuba, Japan, titled "Phosphorus Donors and Boron Acceptors in Silicon Nanowires Synthesized by Laser Ablation." Fukata used a combination of Raman scattering and ESR measurements to show how

P and B dopant incorporate and preferentially segregate in Si nanowires. Dopant control is key to making functional nanowire devices, and the poster by Fukata et al. received the MRS poster award. Another outstanding paper was presented by Irene Goldthorpe, from Stanford, titled "Synthesis and Strain Relaxation Mechanisms of Ge-core/Si-shell Nanowires." The paper, which won the graduate student award silver medal, discussed in detail the strain relief mechanisms in Ge/Si core shell nanowires. Another noteworthy paper which also won the silver medal was presented by Y. Jung, from Materials Science and Engineering, University of Pennsylvania, titled "Phase-Change Nanowires: Size-dependent Electronic Memory Switching and Core/shell Heterostructured Multi-state Memory." Jung et al. reported on two classes of phase change nanowires based on the GST materials system and their memory switching characteristics, and related that these nanowires satisfy many of the technological requirements for successful device implementation.

Semiconductor nanowires are rich in fundamental issues and promise revolutionary new device concepts. Devices fabricated from these nanoscale structures may offer significantly improved photonic and electronic performance. Given the interest, fascination, and rapid development in the field of nanowires, this topic should continue to figure prominently among MRS symposia. Many fundamental issues concerning nanowire growth mechanisms, dopant incorporation, heterostructures, role of surface states, contact formation, and integration into functional devices remain unclear and will undoubtedly generate interest in the scientific community. Equally important are the new opportunities associated with advances in nanowire science and technology, such as applications of these materials in energy harvesting and conversion, chemical and biological sensing, and novel memory and logic devices.

We would like to thank several sponsors for their generous financial support, including the U.S. Office of Naval Research, First Nano Inc. and Hysitron Inc.

Yi Cui
Lincoln Lauhon
A. Alec Talin
E.P.A.M. Bakkers

April 2009

MATERIALS RESEARCH SOCIETY SYMPOSIUM PROCEEDINGS

Volume 1108 — Performance and Reliability of Semiconductor Devices, M. Mastro, J. LaRoche, F. Ren, J-I. Chyi, J. Kim, 2009, ISBN 978-1-60511-080-6
Volume 1109E —Transparent Conductors and Semiconductors for Optoelectronics, J.D. Perkins, T.O. Mason, J.F. Wager, Y. Shigesato, 2009, ISBN 978-1-60511-081-3
Volume 1110E —Theory and Applications of Ferroelectric and Multiferroic Materials, M. Dawber, 2009, ISBN 978-1-60511-082-0
Volume 1111 — Rare-Earth Doping of Advanced Materials for Photonic Applications, V. Dierolf, Y. Fujiwara, U. Hommerich, P. Ruterana, J. Zavada, 2009, ISBN 978-1-60511-083-7
Volume 1112 — Materials and Technologies for 3-D Integration, F. Roozeboom, C. Bower, P. Garrou, M. Koyanagi, P. Ramm, 2009, ISBN 978-1-60511-084-4
Volume 1113E —Low-Cost Solution-Based Deposition of Inorganic Films for Electronic/Photonic Devices, D.B. Mitzi, D. Ginley, B. Smarsly, D.V. Talapin, 2009, ISBN 978-1-60511-085-1
Volume 1114E —Organic and Hybrid Materials for Large-Area Functional Systems, A. Salleo, A.C. Arias, D.M. DeLongchamp, C.R. Kagan, 2009, ISBN 978-1-60511-086-8
Volume 1115 — Physics and Technology of Organic Semiconductor Devices, M. Baldo, A. Kahn, P.W.M. Blom, P. Peumans, 2009, ISBN 978-1-60511-087-5
Volume 1116E —Reliability and Properties of Electronic Devices on Flexible Substrates, J.R. Greer, J. Vlassak, J. Daniel, T. Tsui, 2009, ISBN 978-1-60511-088-2
Volume 1117E —Materials Science for Quantum Information Processing Technologies, M. Fanciulli, J. Martinis, M. Eriksson, 2009, ISBN 978-1-60511-089-9
Volume 1118E —Magnetic Nanostructures by Design, J. Shen, Z. Bandic, S. Sun, J. Shi, 2009, ISBN 978-1-60511-090-5
Volume 1119E —New Materials with High Spin Polarization and Their Applications, C. Felser, A. Gupta, B. Hillebrands, S. Wurmehl, 2009, ISBN 978-1-60511-091-2
Volume 1120E —Energy Harvesting—Molecules and Materials, D.L. Andrews, K.P. Ghiggino, T. Goodson III, A.J. Nozik, 2009, ISBN 978-1-60511-092-9
Volume 1121E —Next-Generation and Nano-Architecture Photovoltaics, V.G. Stoleru, A.G. Norman, N.J. Ekins-Daukes, 2009, ISBN 978-1-60511-093-6
Volume 1122E —Structure/Property Relationships in Fluorite-Derivative Compounds, K.E. Sickafus, A. Navrotsky S.R. Phillpot, 2009, ISBN 978-1-60511-094-3
Volume 1123 — Photovoltaic Materials and Manufacturing Issues, B. Sopori, J. Yang, T. Surek, B. Dimmler, 2009, ISBN 978-1-60511-095-0
Volume 1124 — Scientific Basis for Nuclear Waste Management XXXII, R.B. Rebak, N.C. Hyatt, D.A. Pickett, 2009, ISBN 978-1-60511-096-7
Volume 1125 — Materials for Future Fusion and Fission Technologies, C.C. Fu, A. Kimura, M. Samaras, M. Serrano de Caro, R.E. Stoller, 2009, ISBN 978-1-60511-097-4
Volume 1126 — Solid-State Ionics—2008, E. Traversa, T. Armstrong, K. Eguchi, M.R. Palacin, 2009, ISBN 978-1-60511-098-1
Volume 1127E —Mobile Energy, M.C. Smart, M. Nookala, G. Amaratunga, A. Nathan, 2009, ISBN 978-1-60511-099-8
Volume 1128 — Advanced Intermetallic-Based Alloys for Extreme Environment and Energy Applications, M. Palm, Y-H. He, B.P. Bewlay, M. Takeyama, J.M.K. Wiezorek, 2009, ISBN 978-1-60511-100-1
Volume 1129 — Materials and Devices for Smart Systems III, J. Su, L-P. Wang, Y. Furuya, S. Trolier-McKinstry, J. Leng, 2009, ISBN 978-1-60511-101-8
Volume 1130E —Computational Materials Design via Multiscale Modeling, H.E. Fang, Y. Qi, N. Reynolds, Z-K. Liu, 2009, ISBN 978-1-60511-102-5
Volume 1131E —Biomineral Interfaces—From Experiment to Theory, J.H. Harding, J.A. Elliott, J.S. Evans, 2009, ISBN 978-1-60511-103-2

MATERIALS RESEARCH SOCIETY SYMPOSIUM PROCEEDINGS

Volume 1132E —Mechanics of Biological and Biomedical Materials, R. Narayan, K. Katti, C. Hellmich, U.G.K. Wegst, 2009, ISBN 978-1-60511-104-9
Volume 1133E —Materials for Optical Sensors in Biomedical Applications, D. Nolte, P. Kiesel, X. Fan, G. Hong, 2009, ISBN 978-1-60511-105-6
Volume 1134 — Polymer-Based Smart Materials—Processes, Properties and Application, Z. Cheng, Q. Zhang, S. Bauer, D.A. Wrobleski, 2009, ISBN 978-1-60511-106-3
Volume 1135E —Design, Fabrication, and Self Assembly of "Patchy" and Anisometric Particles, E. Luijten, S.C. Glotzer, F. Sciortino, 2009, ISBN 978-1-60511-107-0
Volume 1136E —Materials in Tissue Engineering, T. Webster, 2009, ISBN 978-1-60511-108-7
Volume 1137E —Nano- and Microscale Materials—Mechanical Properties and Behavior under Extreme Environments, A. Misra, T.J. Balk. H. Huang, M.J. Caturla, C. Eberl, 2009, ISBN 978-1-60511-109-4
Volume 1138E —Nanofunctional Materials, Structures and Devices for Biomedical Applications, L. Nagahara, T. Thundat, S. Bhatia, A. Boisen, K. Kataoka, 2009, ISBN 978-1-60511-110-0
Volume 1139 — Microelectromechanical Systems—Materials and Devices II, S.M. Spearing, S. Vengallatore, J. Bagdahn, N. Sheppard, 2009, ISBN 978-1-60511-111-7
Volume 1140E —Advances in Material Design for Regenerative Medicine, Drug Delivery and Targeting/Imaging, V.P. Shastri, A. Lendlein, L.S. Liu, S. Mitragotri, A. Mikos, 2009, ISBN 978-1-60511-112-4
Volume 1141E —Bio-Inspired Transduction, Fundamentals and Applications, T. Vo-Dinh, C. Liu, A. Zribi, Y. Zhao, 2009, ISBN 978-1-60511-113-1
Volume 1142 — Nanotubes, Nanowires, Nanobelts and Nanocoils—Promise, Expectations and Status, P. Bandaru, S. Grego, I. Kinloch, 2009, ISBN 978-1-60511-114-8
Volume 1143E —Transport Properties in Polymer Nanocomposites, J. Grunlan, M. Ellsworth, S. Nazarenko, J-F. Feller, B. Pivovar, 2009, ISBN 978-1-60511-115-5
Volume 1144 — Nanowires—Synthesis, Properties, Assembly and Applications, Y. Cui, E.P.A.M. Bakkers, L. Lauhon, A. Talin, 2009, ISBN 978-1-60511-116-2
Volume 1145E —Applications of Group IV Semiconductor Nanostructures, T. van Buuren, L. Tsybeskov, S. Fukatsu, L. Dal Negro, F. Gourbilleau, 2009, ISBN 978-1-60511-117-9
Volume 1146E —*In Situ* Studies across Spatial and Temporal Scales for Nanoscience and Technology, S. Kodambaka, G. Rijnders, A. Petford-Long, A. Minor, S. Helveg, A. Ziegler, 2009, ISBN 978-1-60511-118-6
Volume 1147E —Grazing-Incidence Small-Angle X-Ray Scattering, B. Ocko, J. Wang, K. Ludwig, T.P. Russell, 2009, ISBN 978-1-60511-119-3
Volume 1148E —Solid-State Chemistry of Inorganic Materials VII, P.M. Woodward, J.F. Mitchell, S.L. Brock, J.S.O. Evans, 2009, ISBN 978-1-60511-120-9
Volume 1149E —Synthesis and Processing of Organic and Polymeric Functional Materials for a Sustainable Energy Economy, J. Li, C-C. Wu, S.Y. Park, F.B. McCormick, 2009, ISBN 978-1-60511-121-6
Volume 1150E —Artificially Induced Grain Alignment in Thin Films, V. Matias, R. Hammond, S-H. Moon, R. Hühne, 2009, ISBN 978-1-60511-122-3
Volume 1151E —Selecting and Qualifying New Materials for Use in Regulated Industries, R. Rogge, J. Theaker, C. Hubbard, R. Schneider, 2009, ISBN 978-1-60511-123-0
Volume 1152E —Local Structure and Dynamics in Amorphous Systems, Jeff Th.M. de Hosson, A.L. Greer, C.A. Volkert, K.F. Kelton, 2009, ISBN 978-1-60511-124-7

Prior Materials Research Society Symposium Proceedings available by contacting Materials Research Society

Mater. Res. Soc. Symp. Proc. Vol. 1144 © 2009 Materials Research Society 1144-LL01-04

Growth of Ultra Thin ZnSe Nanowires

Tai-Lun Wong, Yuan Cai, Siu-Keung, Chan, Iam-Keong Sou and Ning Wang

Department of Physics and the William Mong Institue of Nano Science and Technology, The Hong Kong University of Science and Technology, Hong Kong, China

ABSTRACT

We report here the growth of ultra thin ZnSe nanowires at low temperatures by Au-catalyzed molecule beam epitaxy and structural characterization of the nanowires. ZnSe nanowires may contain a high density of stacking faults and twins from low temperature growth and show a phase change from cubic to hexagonal structures. Ultra thin ZnSe nanowires can grow at a temperature below the eutectic point, and the relationship between the growth rates and nanowire diameters is V= $1/d^n + C_0$ (C_0 is a constant and n is a fitting parameter). The growth rate of the ultra thin nanowires at low temperatures can be elucidated based on the model involving interface incorporation and diffusion, in which the catalyst is solidified, and the nanowire growth is controlled through the diffusion of atoms into the interface between the catalyst and nanowire. The growth rate of ZnSe ultra thin nanowires has been simulated.

INTRODUCTION

In the classical vapor-liquid-solid (VLS) model [1,2], it is believed that the metal catalyst is in molten state which absorbs the source materials to form a supersaturated liquid droplet. The precipitation of the source atoms occurs at the droplet-whisker interface, and the precipitation rate is mainly determined by the supersaturation of the droplet. Givargizov *et al.* [1,2] determined the whisker growth rate as a function of the driving force of supersaturation ($\Delta\mu/kT$) and first empirically described the growth rate by

$$V = \frac{dL}{dt} = b\left(\frac{\Delta\mu_o}{k_B T} - \frac{4\Omega\sigma}{dk_B T}\right)^n \quad (1),$$

where d is the nanowire diameter, T is the growth temperature, k_B is Boltzmann's constant $\Delta\mu_0$ is the effective difference between the chemical potentials of source element in the nutrient phase and in solid phase, Ω is the atomic volume source element, σ is the specific of the nanowire surface, *b* and *n* (~2) were empirical fitting parameters, . According to Equation (1), the larger the whisker diameter, the faster is its growth rate. This growth phenomenon is attributed to the well-known Gibbs-Thomson effect, i.e., the decrease of supersaturation as a function of the whisker diameter. [1, 2]

Due to the change of the driving force, Si whiskers with small diameters (< 100 nm) grow very slowly. Obviously, there is a critical diameter at which $\Delta\mu = 0$ and the whisker growth stops completely. Those whiskers with diameters less than the critical diameter (about 50nm) should stop growing. However, in recent years, both experimentalists and theorists [4-6] have showed that semiconductor nanowires with diameter smaller than 50 nm can grow and show interesting growth behaviors. For example, in the growth of thin

Si and ZnSe nanowires catalyzed by Au particles, thinner nanowires grow faster than thicker ones,[5,6] and most ultra thin nanowires grow at relatively low temperatures. Recently, using *in-situ* transmission electron microscopy (TEM), Kodambaka *et al.* [6] demonstrated that solid catalysts led to Ge nanowire growth even at a temperature below the eutectic point. In this paper, we present the structural changes of these nanowires grown at low temperatures by molecular-beam epitaxy (MBE) and their interesting growth behavior, which is totally different from that predicted by the classical VLS mechanism [1,2] or other growth models controlled by surface incorporation and diffusion mechanisms.[7]

EXPERIMENTAL DETAILS

ZnSe nanowires were grown by a VG V80H MBE system which was dedicated to ZnSe-based II-VI compound growth in a single chamber. A thin Au layer was deposited on a GaAs substrate at 150 °C and then annealed at 530 °C for 10 minutes in order to generate uniform Au nano-catalysts on the substrate surface. ZnSe nanowires were grown at different temperatures using a ZnSe compound source. Details of the experiment setup can be found in a previous paper.[8] The nanowire samples were prepared by cleaving the substrates into small pieces (without any chemical pretreatment) and directly characterized by TEMs (JEOL2010F and Philips CM120).

DISCUSSION

For the VLS growth, the temperature is critical to the nanowire quality and growth direction. [9] As shown in figure 1a, ZnSe nanowires formed at a temperature below 390 °C have non-uniform diameters at the initial growth stage. The nanowire roots are thicker than the tips, and the surfaces near the roots are rough. This is obviously because the deposition of ZnSe on the substrate surface is significant when the temperature is too low and these deposited atoms diffuse from the substrate surface to the nanowire growth fronts. ZnSe nanowires grown at a high temperature do not have this morphology. [8] Once the nanowires reach a certain length, there is no obvious change in the diameters. For the tapered nanowires shown in figure 1a, the diameters may change gradually from 30 nm at the roots to about 5 nm at the tops. The quality of the thick nanowires is poor in comparison with the nanowires grown at a higher temperature (> 530 °C). A high density of stacking faults and twins always occurs in these nanowires. Figure 1b illustrates the stacking faults and twinning structures observed in an individual nanowire. One of the reasons for the formation of these planar defects was due to the phase transformation from the face center cubic (FCC) structure to the hexagonal close-packed (HCP) structure. The chemical composition of the source materials, e.g. the evaporation ratio of Zn:Se and surface energies may also result in the formation of these defects.[10,11] We have observed that ZnSe nanowires formed at a temperature higher than 530 °C were always a cubic structure and contained few defects. Decrease of the growth temperature could result in a high density of stacking faults, nano twins and some portion of HCP structure. However, ultra thin ZnSe nanowires with diameters of about 3-6 nm are always perfect HCP structure at 390 °C (see figure 1c).

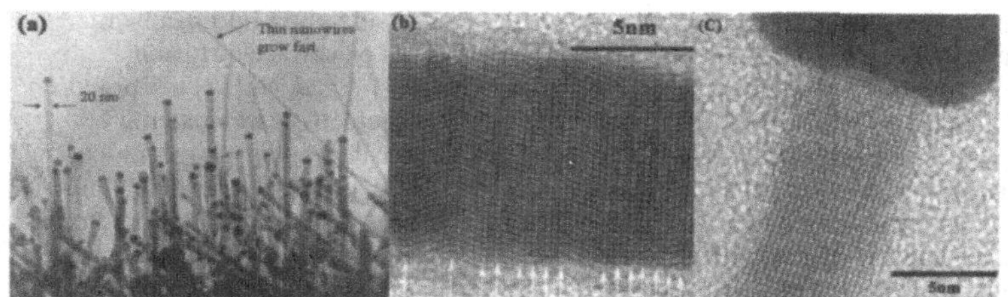

Figure 1(a). ZnSe nanowires with small diameters show fast growth rates. **(b).**High-density stacking faults and twins (marked by the arrows) frequently occur in ZnSe nanowires grown at a low temperature. **(c).** Ultra thin ZnSe nanowires always form HCP structure.

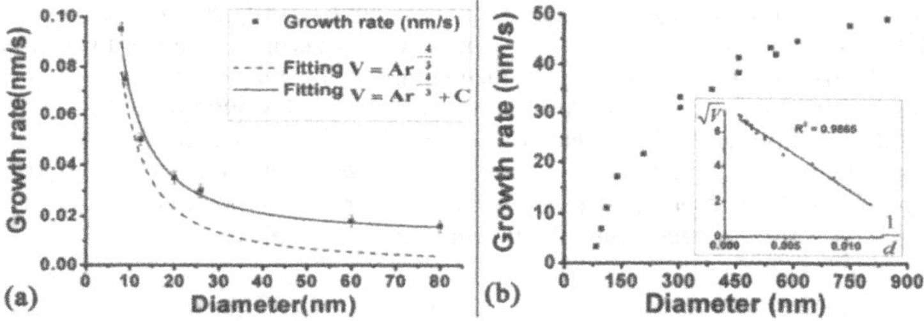

Figure 2(a). The solid dots indicate the growth rates measured from different diameters of ZnSe nanowires. The dashed line is the fitting curve by dL/dt=Ar-4/3 and the solid line by dL/dt=Ar-4/3+C. **(b).** Experimental data of the whisker growth rates reported by Givargizov (1975). [1]

From our measurement, the growth rate of thin ZnSe nanowires (diameters < 60 nm) displays a strong diameter-dependent phenomenon. Smaller nanowires have higher growth rates compared to thicker ones (see figure 2a). This is totally different from the growth of the nanowires or whiskers with diameters greater than 100 nm (see figure 2b, data from Ref. 2). Figure 2a illustrates the changes of growth rates versus the diameters of thin ZnSe nanowires. The relationship between the growth rates and the diameters can be described by $V= 1/d^n + C_0$ (C_0 is a constant and n is about 1-2).This relation does not agree with the classical VLS model,[1,2] in which the metal catalysts are liquid (above the eutectic point), and the nanowire growth is determined mainly by (i) the incorporation of the source atoms on the droplet, (ii) diffusion through the droplet and (iii) precipitation at the liquid-solid interface (see figure 3a). According to our TEM observation, for ultra thin nanowires, however, the growth may largely deviate from these three steps and also deviate from the growth model controlled by surface incorporation and diffusion mechanisms, i.e. the source materials captured by the droplet diffuse along the droplet surface to the growth front (see figure 3b). In this case, the growth rate can be described by $V = 1/r$ [7]. This means that the growth rate is only determined by the circumference of the liquid-solid interface. Similar growth phenomena of diameter-dependence of growth rates have been reported in III-V (e.g., GaAs, GaP, InAs and InP) nanowire growth by different methods, and the growth models based on surface diffusion mechanisms have been proposed. [12,13] For example, Froberg, et al. [12] has proposed a combined model which counts for both the Gibbs-Thomson effect and material diffusion from the substrate surface to explain the diameter-dependent growth rate of InAs nanowires. In their model, for those nanowires with diameters larger than 25nm, it is found that the growth rate is controlled by surface diffusion (the growth rate decreases with increasing the diameter). Due to the Gibbs-Thomson effect, when the nanowire diameters are smaller than 25nm, the growth rate decreases as the diameters shrink. This is because the driving force for the incorporation of atoms to the catalyst is reduced with decreasing the catalyst diameter. In this case, the catalyst is considered as a droplet if the temperature is sufficiently high.

For ultra thin nanowires, however, the growth temperatures are often lower than the eutectic point of the bulk material, and the catalyst may not be a droplet during growth. Due to the Gibbs-Thomson effect, the decrease of the catalyst droplet diameter lowers the solubility of the source atoms and thus shifts the melting temperature of the catalyst. For Au-semiconductor alloys (e.g., Au-Si, Au-Ge and Au-ZnSe), a deviation from the eutectic point generally causes the increase of the melting points. Since the growth of ultra thin nanowires can still maintain under the temperature below the eutectic point (the metal catalyst is solid) as observed by in-situ TEM,[6] the real incorporation and diffusion processes of the source atoms at nanowire tips are complicated.

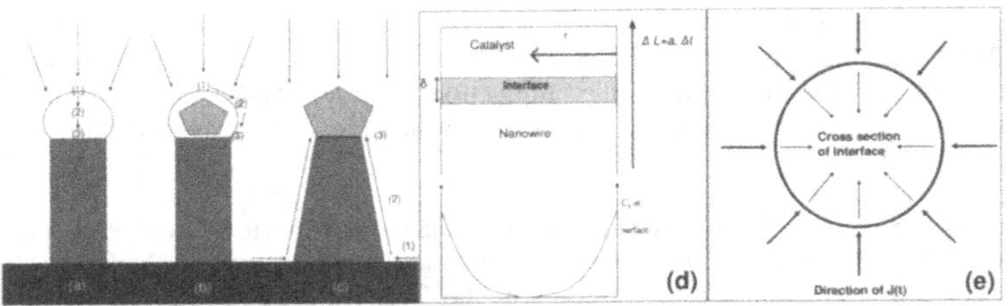

Figure 3. Different diffusion models for the source atoms to incorporate into the growth front of the nanowire. **(a).** The classical VLS. **(b).** The metal droplet is in partially molten state. Its surface and interface are liquid, while the core of the droplet may be solid. **(c).** The metal catalyst is solid, but the interface is liquid. **(d).**The schematic concentration profile of the source atoms at the catalytic interface. **(e).** The flux J(t) of the source atoms flowing into the catalytic interface.

In this paper, we propose a model to interpret the growth behaviors of thin nanowires by solid catalysts. This model is based on (i) the catalyst is solid, (ii) the source atoms deposited on the substrate surface and the nanowire surface diffuse along the nanowire side walls and preferentially flow into the catalyst interface to result in the nanowire growth, and (iii) the nanowire growth rate is mainly due to the atomic diffusion through the interface. Figure 3c shows schematically this surface/interface incorporation process. The interface at the solid catalyst is considered as a grain boundary. Since the grain boundary diffusion is slower than the surface diffusion,[14] the growth of the nanowires in the present model is limited by the interface diffusion. According to the Fisher model for grain boundary diffusion [15], the atom concentration at a grain boundary is described by:

$$c = c_0 \exp(-\pi^{-1/4} \frac{(4D/t)^{1/4}}{(\delta D_{gb})^{1/2}} r) \tag{2}$$

where, D is the volume diffusion coefficient, D_{gb} is the grain boundary diffusion coefficient, δ is the grain boundary width c_0 is the concentration at the surface, r is the radius of the nanowire (see figure 3d). The flux of the atoms flowing into the grain boundary (figure 3e) at time t is:

$$J(t) = -D_{gb}\partial_r c(r,t) \tag{3}$$

where the concentration is quasi-static. Assuming that Δt is the time needed for growing ΔL (one lattice layer of a), then the total number of atoms in ΔL is:

$$N = \pi r_0^2 a\rho = 2\pi r_0 \int_0^{\Delta t} J(t)dt = 2\pi r_0 c_0 \pi^{-1/4} \sqrt{\frac{D_{gb}}{\delta}} (4D)^{1/4} \int_0^{\Delta t} t^{-1/4} dt \tag{4}$$

Then, $$\Delta t = (r_0 \pi^{1/4} \rho \frac{3}{8} \frac{a}{c_0} \sqrt{\frac{\delta}{D_{gb}}} (4D)^{-1/4})^{4/3} \tag{5}$$

and $$\frac{\Delta L}{\Delta t} = \frac{dL}{dt} = \frac{a}{(r_0 \pi^{1/4} \rho \frac{3}{8} \frac{a}{c_0} \sqrt{\frac{\delta}{D_{gb}}} (4D)^{-1/4})^{4/3}} = A \frac{1}{r_0^{4/3}} \tag{6}$$

To estimate the growth rate of ZnSe nanowires versus diameters, we use the following parameters: $D_{gb} = 10^{-12}$ cm^2/s,[14] D= 10^{-16} cm^2/s,[14] $c_0 = 0.95$ $atom/nm^2$;[16] the lattice parameter for ZnSe is a= 0.567 nm; $\rho = 21.7$ $atom/nm^3$. Then, *A* is determined to be 0.5 $nm^{7/3}/s$. Comparing to the experimental data, we obtained:

$$v = A \frac{1}{r_0^{4/3}} + C \tag{7}$$

where, C is about 0.012 nm/s which means the constant deposition of the source atoms. This value is consistent with the deposition rate of ZnSe on a flat substrate at about 390 °C in MBE. The diameter-dependence of the growth rate estimated by equation 7 matches our experimental data fairly well (see figure 2a). We noticed that this interface diffusion mechanism is different from that of the surface diffusion controlled mechanism.[6] This is because the surface diffusivity is generally one order higher than that of the grain boundary diffusivity [14], and for the present model, the deposition of the source atoms is mainly controlled by boundary diffusion. For the nanowires with a diameter d > 60 nm, the growth rate calculated by the present model is very low because of the increase of the diffusion length at the catalyst interface. This dramatically lowers the deposition rate of atoms at the central area of the nanowire-catalyst interface. During the VLS growth, the atoms on the catalyst surface should have a high mobility and are in semi-melting state. Therefore, the surface always can capture the source atoms to cause a constant growth of the nanowire. For ultra thin nanowires, the melting points of the catalysts will decrease quickly by increasing their diameters. For the nanowires with diameters larger than 60 nm, the directly impinging atoms on the liquid catalysts will play a dominant role during the growth, and the nanowire growth will be controlled by the classical VLS mechanism.

CONCLUSIONS

This study shows the structural changes of ultra thin ZnSe nanowires grown at low temperatures, and the interesting diameter-dependent growth behavior of these nanowires. Ultra thin ZnSe nanowires can grow at a temperature below the eutectic point, and their growth rate can be elucidated based on the model involving interface incorporation and diffusion, in which the catalysts are solid. The growth rate of the ultra thin nanowires largely deviates from the classical VLS growth and the simulated relationship between the growth rates and nanowire diameters is $V= 1/d^n + C_0$.

ACKNOWLEDGMENTS

This work was financially supported by the Research Grants Council of Hong Kong (Project Nos. N_HKUST615/06, G_HK021/07,AOE-MG/P-06/06-A) and partially supported by the Nanoscience and Nanotechnology Program at HKUST.

REFERENCES

1. E. I. Givargizov, *J. Cryst. Growth* **31,** 20 (1975).
2. E. I. Givargizov, *Highly Anisotropic Crystal*, (D. Reidel Pub. Co., 1987) pp. 104.
3. T. Y. Tan, N. Li, and U. Gosele, *Applied Physics A-Materials Science & Processing* **78,** 519 (2004).
4. S. Kodambaka, J. Tersoff, M. C. Reuter, and F. M. Ross, *Phys. Rev. Lett.* **96,** 096105 (2006).
5. Y. Cai, S. K. Chan, I. K. Sou, Y. F. Chan, D. S. Su, and N. Wang, *Adv. Mater.* **18**, 109-113 (2006).
6. S. Kodambaka, J. Tersoff, M. C. Reuter, and F. M. Ross, *Science* **316,** 729 (2007).
7. G. Neumann and G. M. Neumann, *Surface Self-Diffusion of Metals, Diffusion Monograph Series*, (Diffusion Information Center, 1972) pp. 105
8. Y. F. Chan, X. F. Duan, S. K. Chan, I. K. Sou, X. X. Zhang, and N. Wang, *Appl. Phys. Lett.* **83,** 2665 (2003).
9. Y. Cai, S. K. Chan, I. K. Sou, Y. T. Chan, D. S. Su, and N. Wang, *Small* **3,** 111 (2007).
10. Z. H. Zhang, F. F. Wang, and X. F. Duan, *J. Cryst. Growth* **303,** 612 (2007).
11. U. Philipose, A. Saxena, H. E. Ruda, P. J. Simpson, Y. Q. Wang, and K. L. Kavanagh, *Nanotechnology* **19,** 215715 (2008).
12. L. E. Froberg, W. Seifert, and J. Johansson, *Phy. Rev. B* **76**, 15340 (2007).
13. W. Seifert, M. Borgstrom, K. Deppert, K. A. Dick, J. Johansson, M. W. Larsson, T. Martensson, N. Skold, C. P. T. Svensson, B. A. Wacaser, L. R. Wallenberg, and L. Samuelson, *J. Cryst. Growth* **272**, 211 (2004).
14. N. A. Gjostein, *Diffusion,* edited by Aaronson, (American Society for Metals, 1973) pp. 241.
15. I. Kaur, *Fundamentals of grain and interphase boundary diffusion*, (John Wiley, 1995) pp. 11,17
16. M. D. Johnson, K. T. Leung, A. Birch, B. G. Orr, J. Tersoff, *Surface Science* **350**, 254-258 (1996)

Mater. Res. Soc. Symp. Proc. Vol. 1144 © 2009 Materials Research Society 1144-LL02-02

Optical Properties of Single Wurtzite GaAs Nanowires and GaAs Nanowires With GaAsSb Inserts

Thang B. Hoang[1], Hailong Zhou[1], Anthonysamy F. Moses[1], Dasa L. Dheeraj[1], A. T. J. van Helvoort[2], Bjørn-Ove Fimland[1] and Helge Weman[1]

[1]Department of Electronics and Telecommunications, Norwegian University of Science and Technology, NO-7491 Trondheim, Norway
[2]Department. of Physics, Norwegian University of Science and Technology, NO-7491 Trondheim, Norway

ABSTRACT

We report results on a low temperature micro-photoluminescence (μ-PL) study of single GaAs nanowires (NWs) and GaAs NWs with GaAsSb inserts. The Au-assisted molecular beam epitaxy (MBE) grown GaAs NWs exhibit wurtzite (WZ) crystal structure with very low stacking fault density. At low temperature (4.4 K), PL emission from single WZ GaAs NWs shows an excitonic band at 1.544 eV, ~25 meV higher in comparison with known zinc-blende (ZB) GaAs band gap energy. The one dimensional heterostructures, GaAs/GaAsSb/GaAs, contain GaAsSb inserts (~20 nm long) inserted in GaAs NWs (~50 nm diameter) and are capped with a radial AlGaAs shell. Due to the type II band alignment at the pseudomorphically strained GaAs/GaAsSb heterojunction, the PL emission observed at ~1.26 eV is believed to be due to the spatially indirect recombination between confined holes in the GaAsSb insert and Coulomb attracted electrons at the GaAs/GaAsSb interfaces. At high excitation power a strong blue shift of the PL energy is observed, characteristic for type II transitions. The realization of WZ GaAs and type II GaAs/GaAsSb core-shell NW heterostructures promise interesting physics as well as potential for developing NW based photonic devices.

INTRODUCTION

Semiconductor NWs have attracted considerable attention because of their potential applications in electronic and opto-electronic devices as well as holding new interesting physical properties.[1-3] III-V semiconductor NWs, such as InP and GaAs NWs, exhibit WZ crystal structure in despite of that most of III-V semiconductor compounds exhibit ZB crystal structure in bulk or thin film form.[4-7] The new WZ crystal structure raises questions about fundamental physical parameters such as band gap energy, exciton binding energy and carrier effective masses of these materials. They also create new possibilities for band-structure engineering in NW based devices, including NW-quantum dot heterostructures which can potentially be used as single photon sources.[8]

In contrast to the case of GaAs NWs grown by metal-organic chemical vapor deposition (MOVCD) where mostly the ZB crystal structure is observed,[9,10] WZ GaAs NWs are normally observed when grown by MBE. Several works have experimentally attempted to examine the band gap and PL emission from WZ GaAs NWs but the results are still inconsistent.[11,12] It is even more challenging to study the optical properties of single NWs, due to the high sensitivity of photo-excited carriers to surface states in GaAs. Another major issue when studying the optical properties of these NWs is the large density of stacking faults which are formed during the growth. The NWs can also contain segments of ZB GaAs which act as collectors that collect most of the photo-excited carriers.[5]

In this work, we report the optical properties of single WZ GaAs NWs. The results on single pseudomorphically strained WZ GaAs NWs with ZB GaAsSb inserts overgrown by a radial AlGaAs shell are also presented as a demonstration of a type II NW heterostructure. In comparison with the known ZB GaAs band gap energy, our low temperature (4.4 K) μ-PL measurements from single WZ GaAs NWs show a PL emission band at an approximately 25 meV higher energy. This result is consistent with previous theoretical calculations.[13,14] PL measurements on single GaAsSb inserts shows a strong emission band at ~1.26 eV. Power dependence measurements show a strong blue shift (~100 meV) of the PL peak, believed to be caused by the state filling effect that induces a renormalization of the ground state energy due to the increasing Coulomb potential as well as from state filling of higher excited holes in the GaAsSb insert.[15]

EXPERIMENTS

The GaAs NWs were grown in a Varian Gen II Modular MBE system equipped with a regular Al cell, a Ga dual filament cell, and Sb and As valved cracker cells. The GaAs(111)B substrate surface was first deoxidized at 620°C, and then a ~ 200 nm thick GaAs buffer layer was grown. This buffer layer was capped with an amorphous As layer to avoid oxidation during its transfer in ambient air to an electron-beam evaporation system for gold depositions. After de-capping, a ~ 1 nm thick Au film was deposited on the sample surface before the sample was loaded into the MBE system again. Under an As_4 flux of 6×10^{-6} Torr, the substrate temperature was increased to a temperature suitable for GaAs NW growth. At this stage, nano-particles containing Au alloyed with the substrate constituents were formed. GaAs NW growth was initiated by opening the shutter of the Ga effusion cell. The temperature of the Ga effusion cell was preset to yield the required Ga flux.

For GaAs/GaAsSb NW heterostructures, GaAs NWs were grown for 25 minutes and then the Sb shutter was opened to supply an additional Sb_2 flux of 6×10^{-7} Torr. A GaAsSb insert was grown for either 30 s, 10 s or 5 s (in different samples), followed by a one minute growth interruption under As_4 flux and then by 10 minutes of GaAs NW growth. The GaAs/GaAsSb NWs were grown under a Ga flux of 4.4×10^{-7} Torr and a substrate temperature of 540°C. In order to suppress non-radiative surface recombination, an AlGaAs shell was grown under As_4 flux with an Al:Ga flux ratio of 3:7. The growth time for the AlGaAs shell was 10 minutes. Finally, a GaAs capping shell was grown (2 minutes) in order to prevent oxidation of the AlGaAs shell.

μ-PL measurements were carried out using an Attocube CFMI optical cryostat. Samples were placed in a He exchange gas with a temperature kept at 4.4 K. Single NWs were excited by a 633 nm HeNe laser line. The laser was defocused onto the samples with an excitation density varied between 0.1-10 $kWcm^{-2}$ using a 0.65 numerical aperture objective lens. The μ-PL from single NWs was collected by the same lens and dispersed by a 0.55 m focal length Jobin-Yvon spectrograph and detected by an Andor-Newton thermo-electric cooled Si CCD camera. The spectral resolution of the system is ~200 μeV. For single NW measurements, NWs were removed from their grown substrate and dispersed on a Si substrate with an average density of ~0.1 NW per μm^2.

DISCUSSION

GaAs nanowires

Figure 1a) shows a transmission electron microscopy (TEM) image of a representative GaAs NW with a low density of stacking faults. In Figure 1b), a high resolution TEM (HRTEM) image of the same NW shows a high quality crystal structure. The diffraction image pattern in the inset of Figure 1b), indicates a pure WZ crystal structure for this NW. μ-PL were measured on single NWs from the same sample from which the TEM images in Figures 1 were taken. Figure 2 shows PL spectra of several single WZ GaAs NWs at 4.4 K. The excitation power was 1 kW/cm^2 and defocused to a spot size of around 5 μm so that individual NWs are entirely excited.

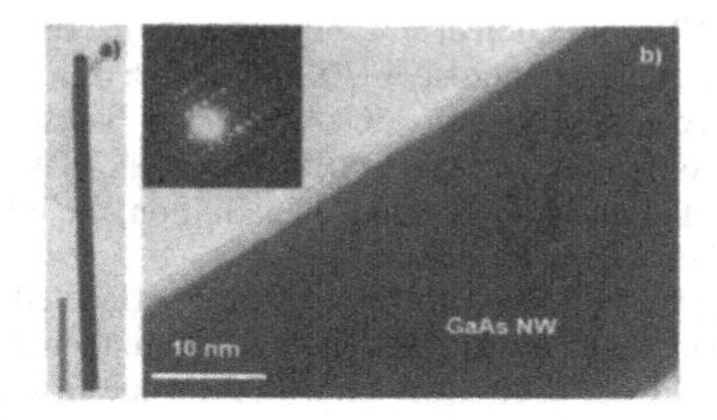

Figure 1. a) TEM image of a stacking fault free GaAs NW. b) HRTEM image and diffraction pattern (inset) shows WZ crystal structure. The scale bar in a) is 500 nm.

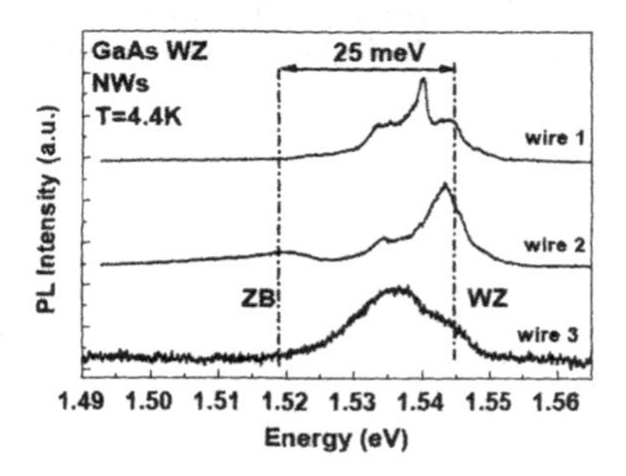

Figure 2. PL spectra from three single NWs. Note that wire 2 shows emissions from both ZB and WZ energies while wires 1 and 3 show only WZ.

The PL emissions from all these NWs exhibit a broad emission band centered at ~1.539eV. The PL spectra display a full width at half maximum (FWHM) of ~11 meV with several features as will be discussed in detail below. All three NWs (wires 1 through 3) show an emission peak near1.545 eV. We suggest this is the free exciton emission of WZ GaAs NWs. This is in consistence with earlier theoretical works[13,14] but in conflict with some recent works (for example, see refs. 11,12,16).

Wire 1 exhibits a narrow emission line at 1.539 eV which might result from excitons bound to defects. The lower energy peaks (between 1.535 and 1.538 eV) which appear in all NWs are probably impurity related recombination peaks. We note that in the PL spectrum of wire 2 there is an emission peak at ~1.518 eV. This emission energy is close to the free exciton emission in ZB GaAs NWs. However, in wires 1 and 3, there are no indications of emission in this energy range. We suggest that this weak emission is associated with ZB GaAs segments in NWs which have a low density of stacking faults. This result is consistent with HRTEM images which show that some NWs exhibit pure WZ free of stacking faults, while other NWs exhibit low density of stacking faults that contain a few ZB GaAs segments.

GaAs nanowires with GaAsSb inserts

As we have reported earlier, WZ GaAs NWs with defect free ZB GaAsSb inserts can be grown expitaxially.[17] Figure 3 shows a scanning electron microscopy (SEM) image of a GaAs/GaAsSb/GaAs NW sample which has been coated with an AlGaAs shell so that the NW heterostructure is pseudomorphically strained. As can be seen in the SEM image, the NWs are straight with a uniform diameter (~ 90 nm including the AlGaAs shell). In addition, the NW density is rather dense and the NW length exhibits a large variation.

In Figure 4a) we plot PL spectra of three GaAs NWs with GaAsSb inserts (labeled A, B and C). As mentioned above, these NWs are coated with an AlGaAs shell therefore their PL intensities are typically two orders of magnitude higher in comparison with the same structure without a shell (not shown here). We note that the PL emission from the GaAs parts of the NWs (Figure 4a) exhibit emission energies close to the ZB band gap energy of GaAs rather than WZ energy as presented in previous section. This is probably due to the presence of stacking faults and ZB GaAs segments contained in this sample. Such ZB GaAs segments are likely to trap most of the photo-excited carriers and therefore emission near the ZB GaAs band gap energy is mostly observed.

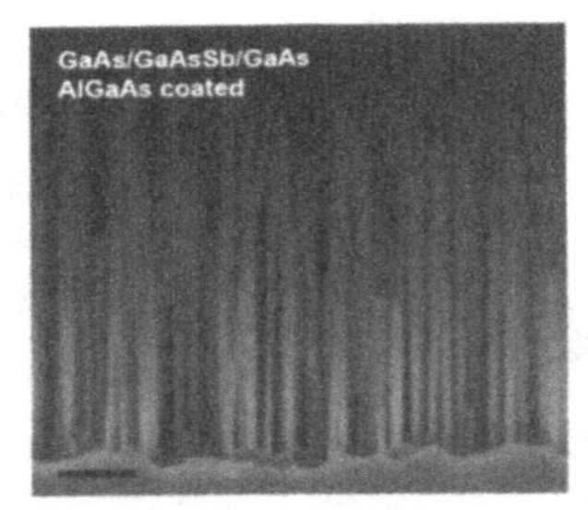

Figure 3. SEM image of a GaAs/GaAsSb/GaAs NW sample (coated with AlGaAs shell). The scale bar is 500 nm.

The emissions from the GaAsSb inserts exhibit a relatively broad band at lower energy (~1.26 eV) and appear to have some variation in energy from wire to wire. This emission is believed to be due to the spatially indirect recombination (type II band alignment) between holes in the GaAsSb insert and Coulomb attracted electrons at the interfaces (see inset in Figure 4b).

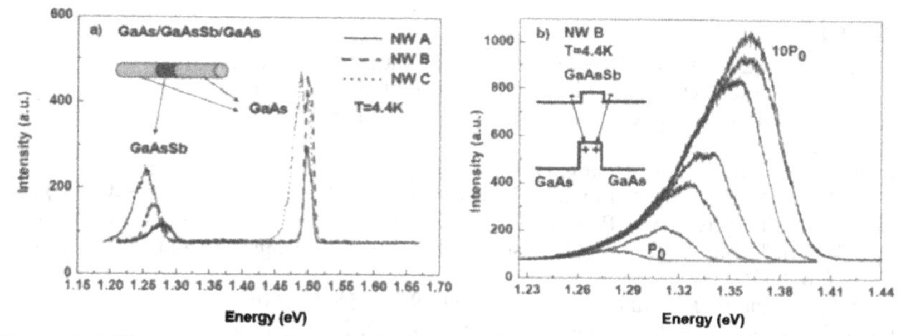

Figure 4 a) PL spectra from three single GaAs NWs (A-C) with GaAsSb inserts. Inset shows a schematic of the heterostructure with arrows indicating from where emissions originate. b) Power dependence of the PL emission from a GaAsSb section of NW B. Inset in b) shows the type II electron-hole recombination in a pesudomorphically strained GaAs/GaAsSb heterostructure.

In Figure 4b) we plot the emission spectra of NW B at difference excitation intensities (P_0=0.1 kW/cm^2). At higher excitation powers, the PL emission becomes broader and strongly blue shifted (by approximately 100 meV). The strong blue shift of the ground state transition is believed to be caused by a renormalization of the ground state energy due to the increasing Coulomb potential, where as the broadening is believed to be due to state filling of higher excited holes in the GaAsSb insert.[15] It is very important to note here that in a similar WZ GaAs NW with a ZB GaAsSb insert structure but without an AlGaAs shell, a type I band alignment could occur due to the relaxation of the strain between GaAs and GaAsSb.[18] A further complication is that the GaAsSb inserts have WZ GaAs barriers, which will also change the conduction and valence band offsets values to the ZB GaAsSb insert.[16] Time-resolved measurements of the exciton lifetimes is probably the best way to clearly distinguish between type I and type II recombination in these structures.

CONCLUSIONS

In conclusion, we have reported the on the optical properties of WZ GaAs NWs and GaAs NWs with GaAsSb inserts. In contrast to bulk or thin film growth, GaAs NWs grown by MBE normally exhibit WZ crystal structure. Low temperature μ-PL measurements on single NWs indicate that the band gap energy of WZ GaAs is ~30 meV higher in comparison with ZB GaAs. We have also investigated the optical properties of pesudomorphically strained GaAs/GaAsSb NWs and observed a strong blue shift in the PL emission energy with excitation power, believed to be a signature of an increasing Coulomb potential characteristic for a type II heterojunction. The realization of WZ-GaAs NWs as well as WZ-GaAs/ZB-GaAsSb NW heterostructures suggest new possibilities for band-structure engineering, which might have potential applications in future NW based photonic devices

ACKNOWLEDGMENTS

This work was supported by the "NANOMAT" program (grant no. 182091) of the Research Council of Norway.

REFERENCES

1. X. Duan, Y. Huang, R. Agarwal, and C. M. Lieber, Nature **421**, 241 (2003).
2. R. Agarwal, C. J. Barrelet, and C. M. Lieber, Nano Lett. **5**, 917 (2005).
3. Y. Gu, E.-S. Kwak, J. L. Lensch, J. E. Allen, T. W. Odom, and L. J. Lauhon, Appl. Phys. Lett. **87**, 043111 (2005).
4. A. Mishra, L. V. Titova, T. B. Hoang, H. E. Jackson, and L. M. Smith, J. M.Yarrison-Rice, Y. Kim, H. J. Joyce, Q. Gao, H. H. Tan, and C. Jagadish Appl. Phys. Lett. **91**, 263104 (2007)
5. H. Pettersson, J. Trägårdh, A. I. Persson, L. Landin, D. Hessman, and L. Samuelson, Nano Lett. **6,** 4 (2006); J. Bao, D. Bell, F. Capasso, T. Mårtensson, J. Trägårdh and L. Samuelson, Nano Lett. **8** 836 (2008)
6. M. I. McMahon and R. J. Nelmes, Phys. Rev. Lett. **95**, 215505 (2005).
7. F. Glas, J.C. Harmand, G. Patriarche, Phys. Rev. Lett. **99**, 146101 (2007).
8. V.I. Klimov et al., Nature **447**, 441 (2007).
9. H. J. Joyce, Qiang Gao, H. Hoe Tan, C. Jagadish, Y. Kim, X. Zhang, Y. Guo, and J. Zou, Nano Lett. **7**, 921-926 (2007).
10. Sköld N, L. S. Karlsson, M. W. Larsson, M-E Pistol, W. Seifert, J. Trägårdh and L. Samuelson, Nano Lett. **5,** 1943-1947 (2005).
11. F. Martelli1, MPiccin, G Bais, F Jabeen, S Ambrosini, S Rubini and A Franciosi, Nanotechnology **18** 125603 (2007).
12. M. Moewe, L. C. Chuang, S. Crankshaw, C. Chase and C. Chang-Hasnain., Appl. Phys. Lett. **93,** 023116 (2008).
13. C-Yu Yeh, S.Z. Lu, S. Froyen and A. Zunger., Phys. Rev. **B46,** 10086 (1992), A. Mujica, R.J. Needs, A. Munoz., Phys. Rev. **B52,** 8881 (1995).
14. M. Murayama and T. Nakayama, Phys. Rev **B49,** 4710 (1994).
15. L. Müller-Kirsch, A. Schliwa, O. Stier, R. Heitz, H. Kirmse, W. Neumann, and D. Bimberg., phys. stat. sol. (b) **224,** 349–352 (2001).
16. Z. Zanolli, F. Fuchs, J. Furthmüller, U. von Barth, and F. Bechstedt, Phys. Rev. **B75,** 245121 (2007).
17. D. L. Dheeraj, G. Patriarche, H. Zhou, T. B. Hoang, A. F. Moses, S. Grønsberg, A. T. J. van Helvoort, B.-O. Fimland and H. Weman., Nano Lett. **8**, 4459 (2008).
18. G. Liu, S-L. Chuang and S-H Park, JAP **88,** 5554 (2000).

Mater. Res. Soc. Symp. Proc. Vol. 1144 © 2009 Materials Research Society 1144-LL03-11

Epitaxial growth of Si nanowires by a modified VLS method using molten Ga as growth assistant

Annika Gewalt[1], Bodo Kalkofen[1], Marco Lisker[1], and Edmund P. Burte[1]
[1] Faculty of Electrical Engineering and Information Technology, Institute of Micro and Sensor Systems, Otto-von-Guericke University,
Magdeburg, Germany

ABSTRACT

In this paper the deposition and morphological characterization of gallium island structures on silicon and first results of silicon wire growth assisted by the created gallium droplets are presented. The islands and wires were grown on (111)-oriented single crystalline p-doped silicon substrates by microwave plasma enhanced chemical vapor deposition (MW PECVD) using trimethylgallium (TMGa) and silane (SiH_4) as precursors for island and wire growth, respectively. The samples were investigated by SEM, EDS, XPS, and AFM.

INTRODUCTION

The epitaxial growth of Si nanowires (NWs) is one of the main subjects among the active research work in semiconductor technology. This is due to the increase of potential applications, and the steadily shrinking of electronic devices according to Moore's law [1]. However, economical utilization of Si NWs in electronic and optoelectronic devices will require the control of the synthesis as well as the reproducible batch growth. In the past decades various methods have been developed for the synthesis of Si NWs. Beside the simply heating of Si wafers to high temperature [2] and the evaporation of Si powders in presence of gold or transition metals as catalysts [3], even several methods for thin film growth have been tested to grow Si whiskers. Under these thin film techniques MBE [4], laser ablation [5], and MOVPE [6] are often mentioned. Most of these techniques are based on the vapor liquid solid (VLS) mechanism, firstly reported by Wagner and Ellis in 1964 [3]. Independent from the process methods used for synthesis, NW growth is very complex. Various possibilities of wire formation have been reported for each technique. Even a slight change of a single process parameter leads to a completely different morphology of the product [7]. Therefore, the particular interests should be the examination of the wires grown under different conditions to reach more knowledge about the growth behavior and the dependencies of the process parameters. Following this ambition, we investigated the synthesis of Si NWs with the assistance of the low melting point metal Ga. In contrast to an Au or Al based VLS mechanism Ga does not work as catalyst, since it does not chemically assist the dissociation of SiH_4, the Si precursor, far below its thermal decomposition point. Gallium and other low melting point metals like indium can only act as a solvent [8, 9]. The decomposition must be enhanced by a plasma treatment [10]. For both, the initial experiments creating the Ga droplets as well as the first attempts in growing Si whiskers, we used MW PECVD. Although PECVD has become one of the most important methods among the various thin film deposition techniques, especially for coating of temperature-sensitive materials, only a few studies previously reported this method as a suitable technique for NW synthesis [9, 10]. According to Hofmann et al. [11], this is because of the simultaneously parasitic deposition of amorphous silicon (a-Si).

EXPERIMENT

The experiments were carried out in a cold wall recipient of a microwave electron resonance (MW-ECR) plasma system called "MicroSys 350" by Roth & Rau.

The process gas was distributed above the substrate via a gas ring inlet. We used pure liquid TMGa precursor as source with H_2 as carrier gas for the initial gallium deposition step. Because of the frequently usage of TMGa as a source material for the fabrication of GaAs semiconductors its behavior is well known on a lot of substrate types. The dissociation of TMGa precursor as well as the atomic arrangement of the Ga adatoms were studied for (100) and (111)-oriented Si by [12-16]. The controlled evaporation of the Ga precursor and the exact mixing with the carrier gas took place in a CEM system of Bronkhorst. The gas line after the vaporization unit can be heated up to 100°C. For the subsequent whisker growth experiments we used SiH_4 as reactant. The deposition support gases (Ar and H_2) were directly let into the plasma source. As substrate we used (111)-oriented single crystalline p-doped Si slices with a diameter of 150 mm.

All the substrates were cleaned for 5 minutes in 120°C hot caro acid followed by a DI water rinse. Then a dip in hydrofluoric acid was carried out for 1 min at room temperature. The wafers were rinsed again and spin dried in nitrogen atmosphere.

Initially, various experiments were carried out to selectively deposit Ga. For this purpose, Si wafer masked with thermally grown and structured SiO_2 were used as substrates. Immediately before the Ga deposition step, these substrates were treated in different ways to remove the native oxide of the unmasked surface. For most of these samples, H_2/Ar plasma treatments were performed in-situ (at 450 °C for 10 min). Other samples were ex-situ cleaned in a buffered oxide etching bath (BHF, at 25°C for 3 min) immediately before growth. The effect of these different treatments on the Ga deposited islands was investigated. Also, the influence of an in-situ annealing procedure after the Ga deposition was analyzed. Therefore, some samples were annealed for 10 min at 600 °C in a forming gas atmosphere.

In the second part of the study, experiments for the growth of Si NWs were carried out. Tab. 1 lists the process parameter set used for the here presented time variation experiments (row 1 is to the Ga deposition step, row 2 stands for Si super saturation).

The morphology of the as-deposited Ga clusters and the as-grown wires were examined by a field emission SEM (S-4800 from Hitachi) as well as by a thermo ionic emission SEM (JSM-5900 from JEOL) typically working with an accelerating voltage in a range of 10-20 kV. Rough chemical composition information about and from the sample surface were investigated by an energy-dispersive X-ray spectroscope (EDS module from RÖNTEC) attached to the JSM-5900. For AFM measurements the Dimension 3100 from Digital Instruments (Veeco) was used to determine the surface roughness, to get information about the topography of the specimens and to carry out statistical analysis of the as-deposited particles. XPS analyses (via ESCA 5600 system with an Al Kα Xray source) were carried out to get chemical information and data with respect to the binding relations from the sample surface and in-depth composition through sputtering with Ar ions.

Tab. 1. Process parameters used for the samples presented in this study (time variation only).

	time [min]	Q_{Ar} [sccm]	Q_{H2} [sccm]	Q_{TMGa} [mg/h]	Q_{SiH4} [sccm]	p [mbar]	U_{DC} [V]	P_{HF} [W]	P_{MW} [W]	T_{Susz} [°C]	T_{CEM} [°C]
1	1-90	10	100	1000	--	0,1	250	50	600	500	50
2	10-60	100	100	--	10	0,2	250	100	600	600	--

DISCUSSION

Gallium deposition on bare and structured substrates

SEM and AFM images revealed that the as-deposited Ga droplets have no preferred orientation. For both pre-growth treatments it was observed that the diameter and the height of the Ga islands increased with deposition time due to the fact that small molten Ga droplets strongly tended to agglomerate [16, 17]. This may be caused by the high surface tension of Ga. Fig. 1 shows some SEM images of islands deposited on BHF cleaned specimens. Fig. 2 illustrates the topographic information, the height signal, and the corresponding deflection amplitude of the sample shown in Fig. 1(c).

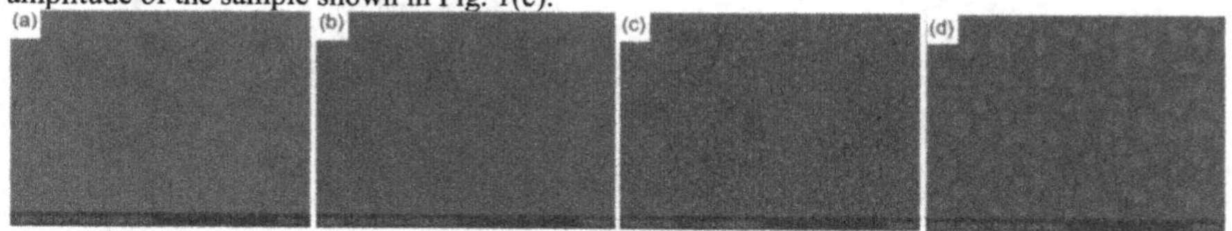

Fig. 1. SEM images showing the arrangement of the as-deposited Ga droplets grown on the previously BHF dipped surfaces without in-situ plasma treatment, processed with different deposition times: (a) 1 min, (b) 3 min, (c) 6 min, and (d) 10 min.

For the ex-situ cleaned samples, the observed islands appeared to be larger than for specimens with identical deposition parameters but treated by the in-situ plasma. We assume, that the plasma can only clean fragments or local areas of the surface by creating craters due to the direct impact of Ar ion bombardment and that the chemical etch assistance of H_2 is too weak to free the surface from native layer in such homogeneous way as the etch dip can. Perhaps, for BHF etched samples, immediately loaded into the reaction chamber, more dangling bonds are available to enable the docking of TMGa and therefore initiate nucleation.

The in-situ annealed samples exhibited a slight increase of the Ga droplet dimensions in contrast to the as-deposited ones. Therefore, it was impossible to get nano-scale dimensions by a subsequent annealing technique as for other VLS metals such as Al or Au [3]. The distribution of the Ga islands still was random after temperature treatment.

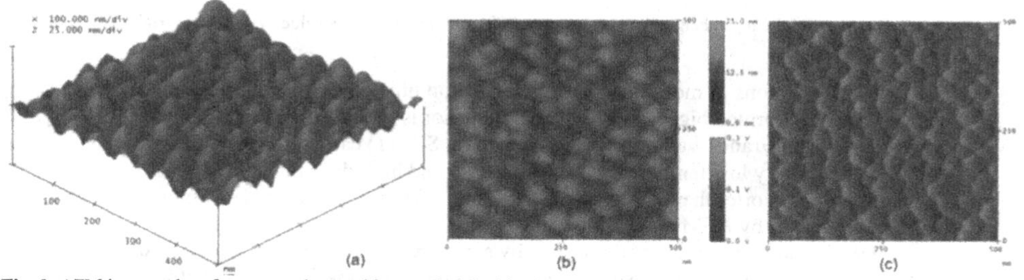

Fig. 2. AFM image taken from a specimen with annealed Ga islands (6 min deposition time at 500 °C and 10 min annealed in forming gas atmosphere at 600 °C) with a scan size of 500 nm; (a) 3D topographical plot, (b) height signal view, and (c) measured deflection voltage.

Statistical analysis showed that the density of the islands decreased with deposition time, which is also explainable by the coalescence of smaller droplets into larger ones and the increasing of the distances between them. The particles also grew in height during a longer TMGa exposure, therefore the root mean square roughness (R_{ms}) of the surface increased with deposition time. The mean values of the dimensions and the R_{ms} of the Ga droplets are summarized together with the duration of deposition in Tab. 2. Fig 3 shows the related curves of the height data and the R_{ms} values versus deposition time.

Tab. 2. Mean island dimensions, roughness and density of annealed Ga islands under different deposition times.

time [s]	h_{mean} [nm]	h_{min} [nm]	h_{max} [nm]	A_{mean} [nm^2]	l_{mean} [nm]	w_{mean} [nm]	$\varnothing_{mean}$ [nm]	R_{ms} [nm]	density [μm^{-2}]
30	1,1	0,41	4,6	844,3	42,6	21,4	28,3	0,6	263,6
45	1,7	0,9	18,8	651,4	37,7	19,7	25,2	1,1	238,7
60	1,9	1,4	16,4	488,1	32,0	18,0	21,9	0,9	145,0
90	1,6	1,1	23,7	592,6	35,5	19,5	24,2	1,0	169,9
120	2,9	1,6	31,5	930,2	42,4	22,6	29,6	1,6	200,9
180	1,5	1,1	22,6	446,9	28,6	18,5	20,3	1,1	122,4
210	3,2	1,0	8,2	1183,3	51,7	24,2	35,0	2,1	263,2
240	4,3	2,5	23,1	832,4	41,5	22,7	29,3	2,9	245,6
360	10,5	5,1	29,3	2063,6	60,6	36,2	46,4	6,9	116,0
480	12,9	5,0	26,1	3200,3	83,9	39,8	58,4	6,6	118,1
600	24,0	15,1	42,6	4105,4	90,8	46,7	66,7	9,0	69,8

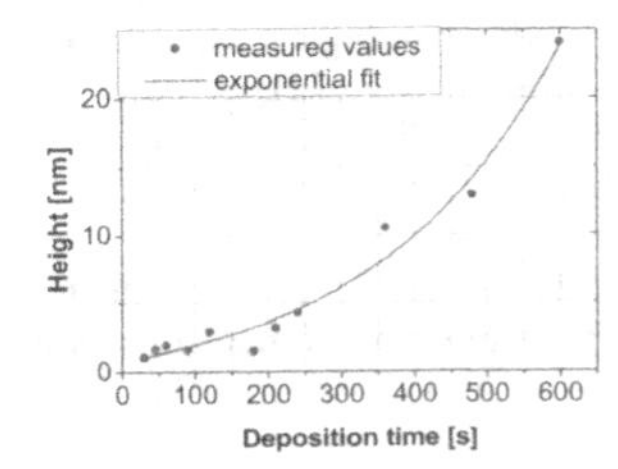

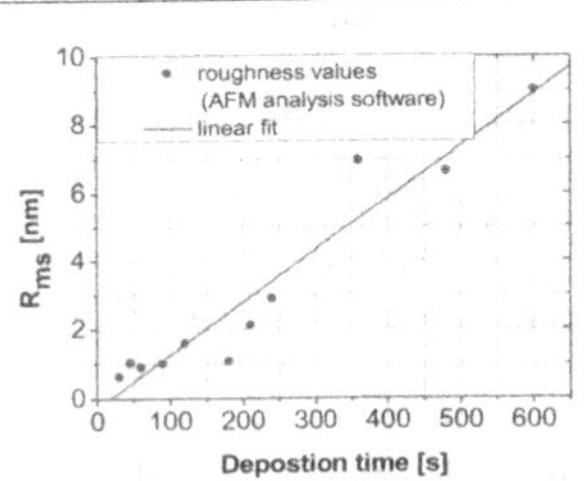

Fig. 3. Diagrams of the mean height data versus time (left) and the roughness values versus time (right) of the in-situ annealed samples.

Ga depositions of more than 60 min deposition time did not lead to a closed monolayer. The islands only formed bigger clusters. This behavior is in agreement with Shogen et al. [17], who reported that Franck-van-der-Merwe growth on Si(111) can only be achieved for a combination of very low temperature (around 80 K) and high dose rates of TMGa. For deposition times shorter than 1 min, no Ga was found on the as-deposited samples, neither by SEM and EDS nor by AFM analyses.

For the samples which were masked by structured SiO_2, it has been observed that the islands only occur on the free Si surface and not on the oxide mask. However, it was found, that the Ga droplets grew in a slightly more ordered manner in arrangement and dimensions along an oxide border (see Fig. 4). Thus, a deposition mask consisting of a fine structured oxide could be used to force a well ordered array of islands on the silicon surface.

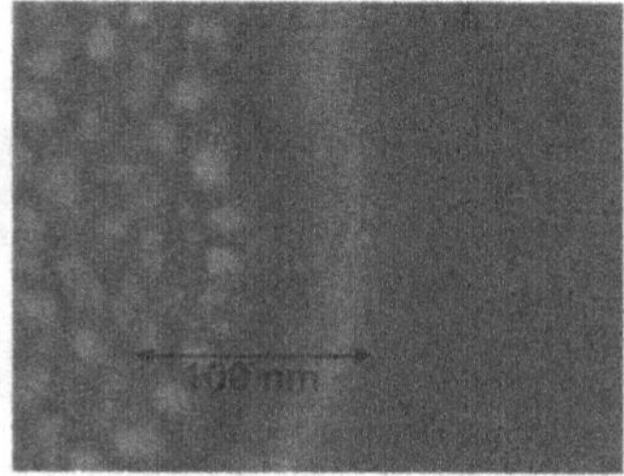

Fig. 4. SEM image of Ga islands deposited onto an oxide masked wafer. An enlargement of the rectangle is shown right. The Ga islands are arranged chain like along the SiO_2 border. Additionally, the dimensions of these droplets are a little more uniformly compared to those islands deposited apart the oxide.

From EDS measurements it has also been observed that the islands only occur on the Si surface and not on the oxide mask. EDS analyses of the unmasked samples showed that the samples were covered by a layer composed of Ga, O, C and an amount of Si. We assume that the detected amount of oxygen may come from a small fraction of native surface oxide grown during the sample transport in air. A XPS survey analysis of one of the in-situ annealed, unmasked samples similarly revealed that the islands were composed of metallic Ga with a non-negligible carbon quantity, possibly coming from the incomplete precursor decomposition [18]. However, the carbon amount in the gallium island stabilizes the metallic Ga phases. Due to the surrounding substrate even Si peaks were identified. XPS analysis after 30 s Ar sputtering showed the presence of elemental Ga, a strongly reduced carbon peak and that the O peak disappeared. This confirms the assumptions done after EDS measurements that the oxygen contamination originated from atmospheric air due to sample transfer. Because the C_{1s} line component was reduced in the spectrum after Ar treatment, we supposed that this peak attributes to both carbon sources: C present on the surface as contaminant and the C incorporated during deposition due to the incomplete decomposition of the precursor. Before Ar bombardment the Ga peaks were small compared to the C and O ones.

Silicon whisker growth – first results

SEM images (Fig. 5) showed differences in the growth types of the synthesized NWs on one and the same sample. Some wires appeared to be relatively uniform, with a smooth surface, similar diameters, and a consistent length. Other seemed to grow bow-like out of the Ga islands and back into the sample surface. These wires had rough surfaces and crossed each other. Even the density of the NWs varied on one specimen, position dependent. Tilted SEM images indicated a random growth direction. However, it was found that many nano-sized wires grew out of a single gallium droplet. This agrees with the classical nucleation theory for critical wire diameters reported elsewhere [6, 9, 10]. Corresponding to Pan et al. [2], the morphologies of the NWs are temperature dependent. Therefore, it could be assumed that the local energy application via the plasma impact could lead to points of higher temperature than on other positions, which are less exposed to plasma and which have a temperature near that of the susceptor. This could explain the discrepancies of the grown wires on the investigated areas of the specimen. However, the substrate temperature is difficult to control independently and is primarily influenced by a combination of MW power and pressure through a convoluted relationship [10]. Even without plasma treatment a temperature difference between substrate holder and wafer surface was observed by experiments at different gas pressures.

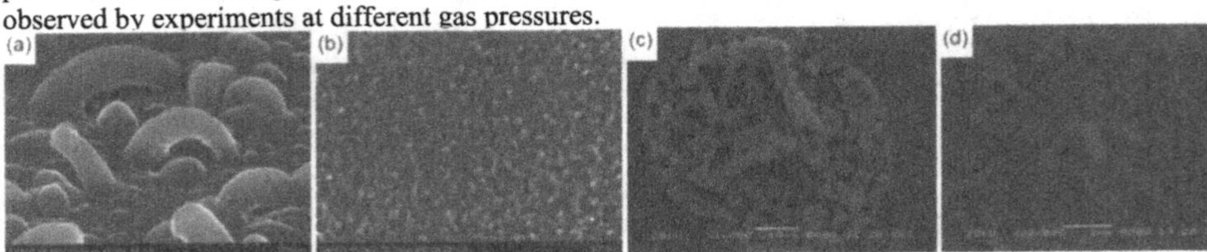

Fig. 5. SEM images of (a) 10 min in-situ plasma treated specimen after 2 min Ga deposition, and 10 min whisker growth reveals a bow-like growth in the central of the sample (60°tilted), (b)grass-like synthesis on a droplet with some distance from the central point of the same specimen (30°tilted), (c) a sample after 60 min whisker growth shows a μm-scale Si structure close to the central part of the specimen (plan view), (d) wool-like growth on a border position on the same sample.

CONCLUSIONS

In the first part of this study the deposition of Ga droplets via PECVD was investigated. It was found, that the Ga islands grew in a disordered manner in arrangement and dimensions. But, a more ordered behavior was found along the borders of an oxide mask. A deposition mask consisting of a fine structured oxide may be used in the future to get a well ordered array of islands on the silicon. Secondly, different NW morphologies have been synthesized via a twice modified VLS process. Our modification used a low melting point metal as solvent instead of a noble metal and PECVD as growth method instead of a conventional CVD technique. SEM observations revealed that the wires formed on the specimen had different sizes, distributions, densities and orientations, which varied with the position on the substrate. It was assumed that the morphologies of the structures were temperature dependent.

Depending on the growth conditions, Ga induces shallow impurities during nanowire growth. To achieve information about the amount of the implemented Ga atoms, electrical measurements are essential. Once the growth of our wires succeeds better, we plan to do electrical characterization of NWs.

REFERENCES

[1] Chen, L. J. (2007). Journal of Materials Chemistry **17**: 4639-4643.
[2] Pan, Z. W., S. Dai, et al. (2003). Nano Letters **3**(9): 1279-1284.
[3] Wagner, R. S. (1970). New York, Wiley.
[4] Schubert, L., P. Werner, et al. (2004). Applied Physics Letters **84**(24): 4968-4970.
[5] Appell, D., P. Yang, et al. (2002). Nature Publishing Group **419**: 553-555.
[6] Sacilotti, M., L. Imhoff, et al. (2004). Journal of Crystal Growth **261**: 253-258.
[7] Cai, X. M., A. B. Djurisic, et al. (2005). Journal of Applied Physics **98**: (074313)1-5.
[8] Griffiths, H., C. Xu, et al. (2007). Surface & Coatings Technology **201**: 9215-9220.
[9] Iacopi, F., P. M. Vereecken, et al. (2007). Nanotechnology **18**: 7.
[10] Sharma, S. and M. K. Sunkara (2004). Nanotechnology **15**: 130-134..
[11] Hofmann, S., C. Ducati, et al. (2003). Journal of Applied Physics **94**(9): 6005-6011.
[12] Gow, T. R., R. Lin, et al. (1990). Journal of Crystal Growth **106**: 577-592.
[13] Lüth, H. (1988). J. Vac. Sci. Technol. **A** 7(3): 696-700.
[14] Lee, F., A. L. Backman, et al. (1989). Surface Science **216**: 173-188.
[15] Förster, A. and H. Lüth (1989). J. Vac. Sci. Technol. **B** 7(4): 720-724.
[16] Lin, R. and R. I. Masel (1991). Surface Science **258**: 225-234.
[17] Shogen, S., Y. Matsumi, et al. (1991 Journal of Applied Physics **70**(1): 452-468.
[18] Pan, Z. W., Z. R. Dai, et al. (2002). J. AM. CHEM. SOC. **124**(8): 1817-1822.

Mater. Res. Soc. Symp. Proc. Vol. 1144 © 2009 Materials Research Society 1144-LL03-16

Matrix Formation Leading to Catalyst Free Growth of GaN Nanowires

J. B. Halpern[1]; G. L. Harris[2]; M. He[2]; P. Zhou[2]; C. Cheek[3]
[1]Chemistry, Howard University, Washington, DC 20059, U.S.A.
[2]Howard Nanofabrication Facility, Howard University, Washington, DC 20059, U.S.A.
[3]Electrical Engineering, Howard University, Washington, DC 20059, U.S.A.

ABSTRACT

Catalyst-free vapor-solid GaN nanowire growth occurs when ammonia flows over Ga first forming a GaN matrix, the top layer of which is composed of hexagonal platelets. Multiphase nanowire growth occurs at nanoscale nucleation sites on the GaN platelets. Lower layers of the matrix are Ga rich, upper ones are stoichiometrically GaN. Gallium for later stages of growth is sourced from the decomposition of GaN particles and Ga rich GaN. Growth temperature exerts a strong influence on nucleation site formation. Scanning electron microscopy (SEM) and Energy Dispersive Spectroscopy (EDS) was used to characterize the matrix.

INTRODUCTION

An increasing number of studies of semiconductor nanowires have appeared recently [1, 2, 3] and many methods have been used to grow them. Characteristically, nanowire growth by chemical vapor deposition (CVD) occurs at lower temperatures than thin film growth and the nanowire growth occurs along preferred axes. This requires that the limiting activation energy for decomposition of the CVD source(s) be exceptionally low at the surface along which the nanowire grows. In other words, nanowire growth selects an orientation along which growth is kinetically favored. The decomposition can be either catalytic or not. Information on the most favored orientations for crystal growth could be useful for thin film growth. Catalytic GaN nanowire growth involves catalytic vapor-liquid-solid formation. [4] In that case, liquid metal droplets, usually Ni, decompose ammonia and solvate Ga metal vapor. Growth of wurtzite GaN wires oriented along the c axis then occurs at the bottom of the droplet.

This paper describes a vapor-solid mechanism for catalyst free growth of GaN nanowires by reaction of Ga metal vapor with NH_3 in a tube furnace. [5, 6, 7, 8] The wires can either be wurtzite, or biphasic combinations of wurtzite and zinc-blende. [9, 10, 11] The structure of the wire remains constant along its entire length. Growth temperature determines the form.

Initially, a Ga rich matrix of GaN appears. In a short time the top of the matrix becomes crystalline and stoichiometrically pure GaN. Small hexagonal platelets appear on the surface. Wires grow from the edge and the center of the platelets. Their thickness (20 nm to 10 μ) can be controlled by varying the ammonia flow rate (20-150 sccm) and temperature (800-1100 °C). Wires that grow from the edge of the platelets are as thick as the platelets and are biphasic. Wires that grow from the centers of the platelets (along the c-axis in the [0001] direction) are pure wurtzite and generally thicker. Length is determined by the duration of the growth period.

EXPERIMENT

Roughly 3 g of Ga metal in a small boron nitride boat was set in the middle of a short, 300 mm long, 20 mm diameter quartz tube. This liner was inserted into a 25 mm diameter process tube which rested in a single zone tube furnace. Ammonia was introduced into the quartz tube through a mass-flow controller at rates of 20–140 sccm. During growth, the pressure at the upstream end of the process tube was kept between 10 and 20 Torr as measured by a capacitance manometer. The temperature for each run was set between 850 and 1000 °C. Ramp up time from room temperature to the operating temperature took roughly 15 minutes. Cool down requires more than an hour. During cool down, ammonia continued to flow until the temperature reached 500 °C at which point it was replaced by a flow of nitrogen.

RESULTS

Figure 1a shows the quartz liner after 15 minutes growth at 900 °C with an ammonia flow rate of 100 sccm. The pressure at the upstream end of the process tube was 10 Torr. The gray mark at the right was underneath the boat, showing how the Ga spreads initially. If nitrogen is used as carrier gas or the temperature is below 700 °C the Ga metal does not spread but splatters and beads up. Figure 1b is a 20X image of the gray material. The 50X insert shows how it is striated. Figure 1c shows a 50X cross-section. One sees lines of shiny Ga metal in Fig 1b

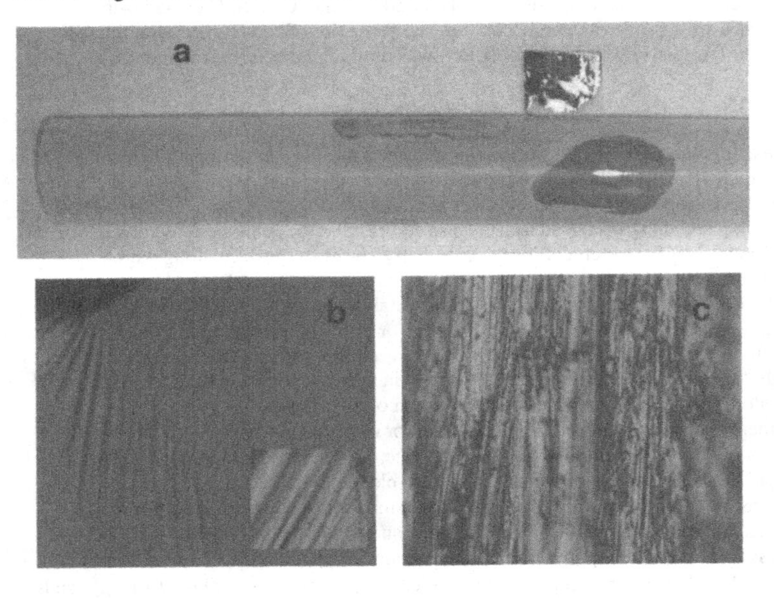

Figure 1. The quartz liner after 15 minutes at 1000 °C a) with a fragment of the BN boat above. The liner has a yellowish cast from absorption of Ga vapor on oxygen atoms in the quartz liner surface forming GaO. The gray mark at the right is GaN formed where the BN boat was. b) is a 20X optical image of the surface of this grayish material, the insert is 50X and c) is a 50X image of the cross-section

and 1c. The material appears layered with alternating Ga metal and GaN. Enough Ga has been converted to GaN to form a very Ga rich matrix.

Fig. 2 shows the matrix grown for 30 and 60 minutes with EDS analysis at three points at different levels. The deeper the layer, the more Ga rich; the longer the growth period the more it is stoichiometrically GaN. Small crystals of about 1 μ diameter appear on the surface, with uniform thickness. Nanowires grow from some of these. The small hexagonal crystal platelets

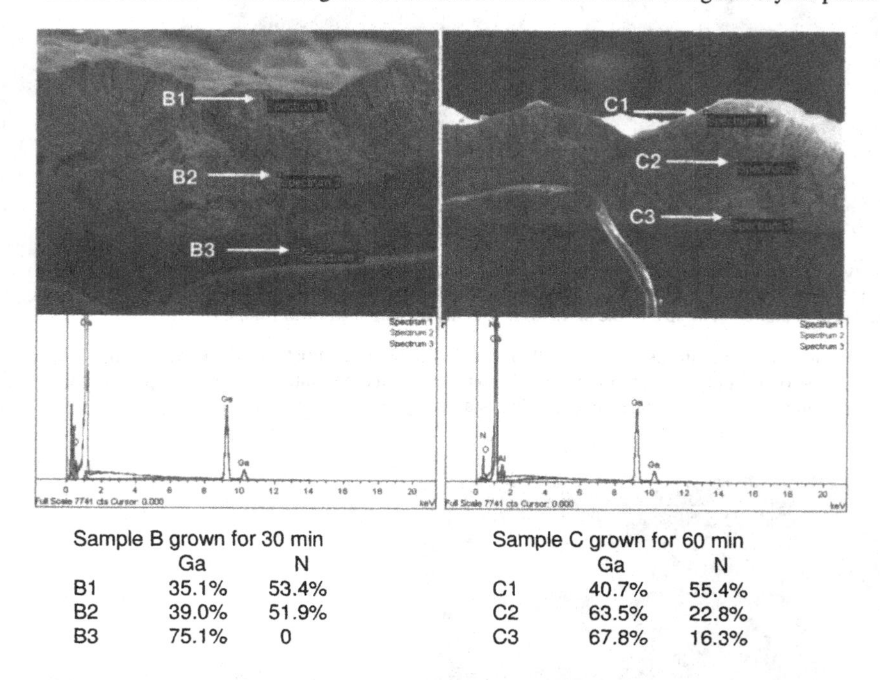

Sample B grown for 30 min

	Ga	N
B1	35.1%	53.4%
B2	39.0%	51.9%
B3	75.1%	0

Sample C grown for 60 min

	Ga	N
C1	40.7%	55.4%
C2	63.5%	22.8%
C3	67.8%	16.3%

Figure 2. Two samples of the GaN matrix were analyzed by EDS. The sample to the left was grown at 1000 °C for 30 minutes, the one on the right for 60 minutes.

are packed tightly together immediately below the surface. As one moves further down in the matrix the small crystals lose definition and corners look like melted candle wax. There is more space between these particles with random, rounded edges, many of which have at least one 120^0 sharp corner which later could develop into a corner of a hexagonal GaN crystal. The appearance of the lower layer suggests that at least a portion of these particles were liquid that later solidified in the cool down. Also, as the growth period lengthens, the round cornered GaN particles become rarer and the sharp edged hexagonal crystals increase in number and size.

Fig. 3 shows the crystalline nature of the upper part of the matrix (left) and the more amorphous lower layer (right). The arrow in Fig. 3a points to a typical platelet. After 30 minutes, no free Ga can be found, therefore the Ga source for nanowire growth must be decomposition of the platelets or of the underlying Ga rich material.

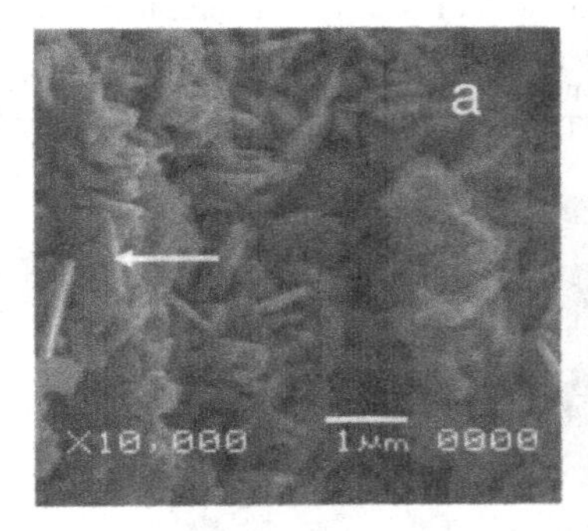

Figure 3. a) An SEM image of the top layer of the matrix showing platelets with sharp corners. b) The lower layer where the solidified particles have random shapes and a rounded appearance. The size of the particles is roughly the same in both images

At lower temperatures, wire growth at the surface occurs mostly at the edges of the platelets as shown in Fig. 4a, with the wires being as thick as the platelet. The flat tip of a typical nanowire can be seen in 4b. The platelet edge is actually complex, having many layers. [8]

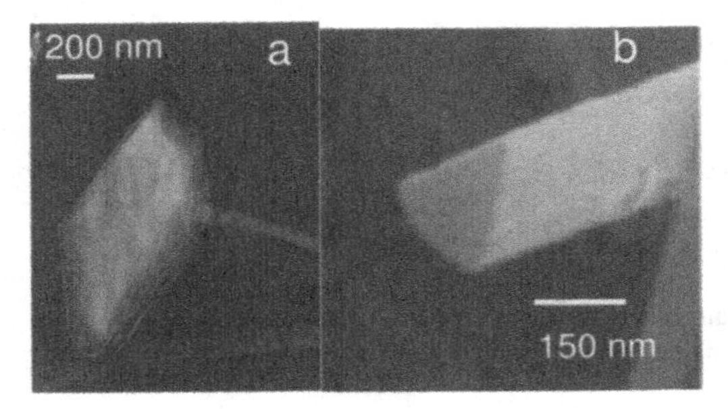

Figure 4. TEM pictures of a) A typical wire grown at low temperature emerging from the corner of a platelet. b) The end of another, larger nanowire extending from a corner of a platelet grown at a relatively low temperature, 850 °C

No Ga is observed at the tips of the nanowires in the catalyst free growth [8], although liquid catalyst is definitely found on the tips for growth catalyzed by Ni and other metals.[3] This is contrary to the assertions of Mohammad [12] for the same catalyst free process described here. Although some free Ga may be trapped in the lower levels of the matrix as discussed above (Fig. 3b), it only serves as a source of vapor phase Ga. We thus conclude that non-catalytic GaN nanowire growth is a vapor-solid process.

DISCUSSION

To summarize, the initial spread of Ga metal liquid requires the presence of gaseous ammonia otherwise Ga metal beads up on the BN boat or the quartz liner surface and does not wet the surfaces. Conversion of the top layer of the Ga to GaN is rapid, within 30 minutes at 900 °C and 100 sccm NH_3. This conversion takes more time at lower temperatures and ammonia flow rates and shorter time at higher temperatures and flow rates. GaN particles that form initially are flat with generally bulbous edges but some edges are at the 120° angle characteristic of wurtzite GaN crystals. The surface is always more crystalline than the lower layers. Nanowires grow out of platelets on the surface of the matrix.

With increasing time and ammonia flow, the size of the crystals on the surface becomes larger and the surface lumpier. This produces a rougher surface. After the first 15 minutes, the crystals on the surface are always hexagonal and show no evidence of having been liquid at the growth temperature as those in the lower GaN layer do. The thickness of the crystals on the surface does not change much with growth time but the basal plane extent does increase.

The detailed process for molecular beam epitaxy (MBE) thin film GaN growth was first described by Wickenden, et al. [13] As temperature increases, wurtzite hexagonal GaN (*h*-GaN) platelets form and provide nuclei for thin film. The initially formed grey GaN matrix appears amorphous, but there are very small crystalline domains buried within as can be seen from x-ray diffraction. This has also been observed for catalyst free GaN nanowire growth. A significant difference is that for MBE growth of thin films, crystals in the nucleation layer which form at very low temperatures are mostly cubic. Their crystallinity develops rapidly as temperatures are raised and the cubic GaN (*c*-GaN) transforms to the hexagonal wurtzite phase. [14]

The first stage of nanowire growth, formation of the amorphous GaN matrix, is analogous to formation of GaN nucleation layers on substrates for molecular beam epitaxial (MBE) growth of thin films. The second stage of nanowire synthesis, growth of GaN platelets, is equivalent to the formation of hexagonal GaN nuclei in thin film growth when the nucleation layer is annealed. Catalyst free nanowire growth has been carried out at a single temperature between 800 °C, and 1000 °C. On the other hand, MBE grown nucleation layer formation usually starts at 500-600 °C where the GaN matrix is created, followed by rapid ramping to temperatures > 800 °C. We have attempted to grow GaN nanowires by forming a nucleation layer at the lower temperatures, but no matrix forms.

The source of Ga atoms for nanowire growth appears to be decomposition of the Ga/GaN matrix. Koleske, et al. studied the conversion process in thin films. [15, 16, 17] Ga from GaN can desorb producing a high local Ga vapor density that can be incorporated into nanoscopic *h*-GaN nuclei or nanowires. [18] Physisorbed NH_3 simply desorbs. Surface incorporated nitrogen atoms can recombine to form N_2. Decomposition of the nucleation layer starts at ~ 800 °C for GaN on sapphire, in good accord with where GaN nanowire growth starts. Compared to GaN thin films, nanowires are grown at low temperatures, total pressures and ammonia flow rates.

CONCLUSIONS

The early stages of catalyst free GaN nanowire growth have been described. The nanowire formation mechanism has been shown to be vapor-solid as opposed to the vapor-liquid-solid mechanism for metal catalyzed growth

ACKNOWLEDGMENTS

Research supported by the U.S. National Science Foundation, Division of Materials Sciences under Grant No. DMR-0611595. Figure 4a is from Reference 8, Figure 4b from Reference 11 both by permission.

REFERENCES

[1] H.J. Fan, P. Werner and M. Zacharias, *Small*, **2**, 700 (2006)
[2] R. Agarwal and M. Lieber Charles, *Applied Physics A: Materials Science & Processing*, **85**, 209 (2006)
[3] M. Law, J. Goldberger and P. Yang, *Annual Review of Materials Research*, **34**, 83 (2004)
[4] C.-C. Chen, C.-C. Yeh, C.-H. Chen, M.-Y. Yu, H.-L. Liu, J.-J. Wu, K.-H. Chen, L.-C. Chen, J.-Y. Peng and Y.-F. Chen, *Journal of the American Chemical Society*, **123**, 2791 (2001)
[5] M.Q. He, I. Minus, P.Z. Zhou, S.N. Mohammed, J.B. Halpern, R. Jacobs, W.L. Sarney, L. Salamanca-Riba and R.D. Vispute, *Applied Physics Letters*, **77**, 3731 (2000)
[6] R.N. Jacobs, L. Salamanca-Riba, M.Q. He, G.L. Harris, P. Zhou, S.N. Mohammad and J.B. Halpern, *Materials Research Society Symposium Proceedings*, **675(Nanotubes, Fullerenes, Nanostructured and Disordered Carbon)**, W9.4.1 (2001)
[7] S.N. Mohammad, G.L. Harris, J.B. Halpern, R. Jacobs, S. W.L., L. Salamanca-Riba, M.Q. He and P.Z. Zhou, *Journal of Crystal Growth*, **231**, 357 (2001)
[8] J.B. Halpern, A. Bello, J. Gilcrease, G.L. Harris and M. He, *Microelectronics Journal*, doi: 10.1016/j.mejo.2008.07.022 (2008)
[9] V.M. Ayres, B.W. Jacobs, M.E. Englund, E.H. Carey, M.A. Crimp, R.M. Ronningen, A.F. Zeller, J.B. Halpern, M.Q. He, G.L. Harris, D. Liu, H.C. Shaw and M.P. Petkov, *Diamond and Related Materials*, **15**, 1117 (2006)
[10] B.W. Jacobs, V.M. Ayres, M.P. Petkov, J.B. Halpern, M. He, A.D. Baczewski, K. Mcelroy, M.A. Crimp, J. Zhang and H.C. Shaw, *Nano Letters*, **7**, 1435 (2007)
[11] B.W. Jacobs, V. Ayers, R.E. Stallcup, A. Hartman, M.A. Tupta, A. Baczewski, D., M. Crimp, A., J. Halpern, B., M. He and H. Shaw, C., *Nanotechnology*, **18**, 475710 (2007)
[12] M. He, A. Motayed and S.N. Mohammad, *Journal of Chemical Physics*, **126**, 064704 (2007)
[13] A.E. Wikenden, D.K. Wickenden and T.J. Kietenmacher, *Journal of Applied Physics*, **75**, 5367 (1994)
[14] J. Narayan, P. Pand, A. Chugh, H. Choi and J.C.C. Fan, *Journal of Applied Physics*, **99**, 054313 (2006)
[15] D.D. Koleske, M.E. Coltrin, K.C. Cross, C.C. Mitchell and A.A. Allerman, *Journal of Crystal Growth*, **273**, 86 (2004)
[16] D.D. Koleske, M.E. Coltrin and M.J. Russell, *Journal of Crystal Growth*, **279**, 37 (2005)
[17] D.D. Koleske, M.E. Coltrin, A.A. Allerman, K.C. Cross, C.C. Mitchell and J.J. Figiel, *Applied Physics Letters*, **82**, 1170 (2003)
[18] M. Lada, A.G. Cullis and P.J. Parbrook, *Journal of Crystal Growth*, **258**, 89 (2003)

Mater. Res. Soc. Symp. Proc. Vol. 1144 © 2009 Materials Research Society 1144-LL03-25

Annealing of Nanocrystalline Silicon Micro-Bridges With Electrical Stress

Gokhan Bakan[1], Adam Cywar[1], Cicek Boztug[1], Mustafa B. Akbulut[1], Helena Silva[1] and Ali Gokirmak[1]

[1]Electrical and Computer Engineering, University of Connecticut, 371 Fairfield Way, Storrs, CT 06269, USA

ABSTRACT

Nanocrystalline silicon (nc-Si) micro-bridges are melted and crystallized through Joule heating by applying high-amplitude short duration voltage pulses. Full crystallization of nc-Si bridges is achieved by adjusting the voltage-pulse amplitude and duration. If the applied pulse cannot deliver enough energy to the bridges, only surface texture modification is observed. On the contrary, if the pulse is not terminated after the entire bridge melts, molten silicon diffuses on to the contact pads and the bridge tapers in the middle. Melting of the bridges can be monitored through current-time (I-t) and voltage-time (V-t) measurements during the electrical stress. Conductance of the bridges is enhanced after the electrical stress.

INTRODUCTION

Thin-film transistors (TFTs) are one of the major components in large-area electronics [1]. Imaging and sensor arrays such as active matrix liquid crystal displays and x-ray imagers use TFTs as switches and in peripheral circuitry [1]. Amorphous silicon (a-Si) is commonly used as TFT material for reliable and cost-effective large-area electronics applications. Although a-Si has very low electron mobility [2], it has the advantage of uniform and low-temperature processing which may also provide the opportunity of using flexible materials like plastics as the substrate [3]. Interest in the studies of the silicon crystallization methods has increased in last several decades due to low processing temperature requirements and demand for high performance TFTs[2, 4]. In this work, crystallization of nanocrystalline silicon (nc-Si) bridges by electrical stress [5] is studied as a silicon crystallization method.

Nc-Si films are deposited on a thermally grown oxide in a low-pressure chemical vapor deposition (LPCVD) system with high-level phosphorous doping ($\sim 5 \times 10^{20}$ cm^{-3}) [6]. Nc-films are patterned as wires with large contact pads using photolithography and reactive ion etching (RIE). Nc-Si wires are released from the underlying oxide using buffered oxide etch to form bridges. As fabricated bridges have lengths ranging from 0.5 to 5 μm and widths in the order of 0.5 μm.

EXPERIMENT and DISCUSSION

Large-amplitude short-duration square voltage pulses are applied across the nc-Si bridges by making contact with large nc-Si pads using tungsten probes. Applied voltage and the corresponding current through the bridges are monitored by a high-speed oscilloscope as shown in Figure 1.

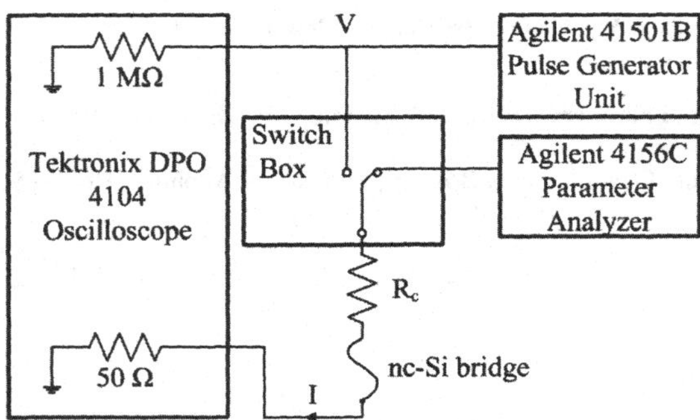

Figure 1. Circuit schematic of the experimental setup, where R_c is the contact-pad resistance. I-V characteristics of the bridges are measured by the parameter analyzer before and after the electrical stress. Switch box switches to pulse generator unit to apply the voltage pulse. Applied voltage pulse amplitude and current through the bridge are acquired by the oscilloscope.

Surface melting and modification (Figure 2b) is observed on the bridges, when the energy delivered by the electrical pulse is not sufficient to melt the entire bridge. Figure 2c shows a melted and crystallized bridge. While the contact pads keep their initial nc-Si structure, the bridge has smooth texture and a lump in the middle. The crystallized bridge is under tensile stress; although as fabricated bridges are relaxed (under compressive stress). The lump in the middle suggests that the solidification process starts from the two ends of the bridge and moves towards the middle. When two fronts meet, excess liquid silicon in the middle is ejected and forms a lump upon solidification. If the electrical stress is not terminated after the melting of the bridge, liquid silicon can diffuse on to the contact pads and the bridge disconnects as shown in Figure 2d. Surface modification in Figure 2b suggests that melting of the bridges starts on the surface.

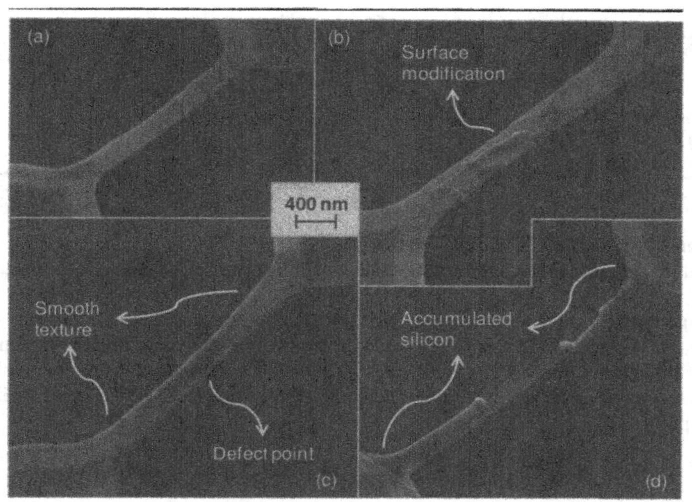

Figure 2. (a) An as fabricated 3.5 μm long nc-Si bridge. The bridge and contact pads have uniform nc-Si texture. (b) A 5 μm long bridge stressed with 30 V, 1 μs pulse. Applied electrical stress melted and modified the surface. (c) A 4 μm long bridge stressed with a 25 V, 1 μs pulse. (d) A 5 μm long bridge stressed with 30 V, 1 μs pulse. The bridge is fully molten and broken.

Melting and re-solidification process can be monitored through the I-t and V-t measurements. In Figure 3, I-t and V-t graphs of the bridges shown in Figure 2b, 2c and 2d are given. In the case of surface modification (Figure 2b), the current increases between the rising and falling edges of the applied pulse without showing any significant jump (Figure 3a). The minimum resistance of the bridge during the stress (15.1 kΩ) is much smaller than its resistance after the stress (76.3 kΩ). Small resistance during the stress can be attributed to the resistivity decrease due to increased temperature and negative temperature coefficient of resistivity [7] of nc-Si, and partial (surface) melting of the bridge. In the case of complete crystallization (Figure 2c), I-t behavior shows a sudden jump (indicated with the dashed line in Figure 3b) during the stress. Gradual increase in the current after the drastic jump suggests that melting of the bridge continues during the entire pulse. If the entire bridge melts before the pulse is terminated, molten silicon diffuses on to the contact pads (Figure 2d) leads to breaking of the bridge for sufficiently long pulses (Figure 3c).

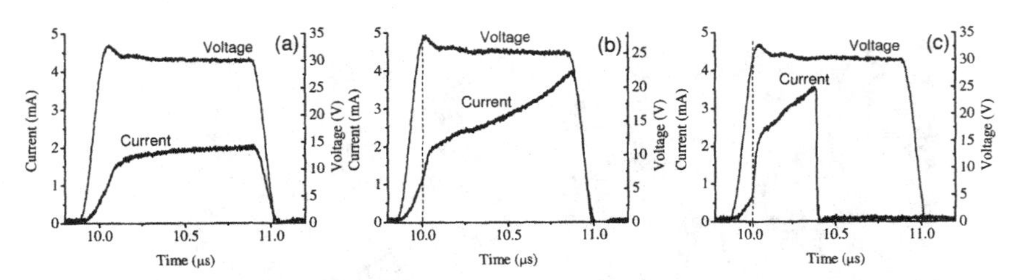

Figure 3. V-t and I-t graphs of the bridges (a) in Figure 2b, (b) in Figure 2c, (c) Figure 2d. Vertical dashed lines in (b) and (c) indicate the time when the current drastically increases.

I-V characteristics and conductance of the crystallized bridge before and after the electrical stress are shown in Figure 4. Conductance versus voltage behavior shows that tungsten probes are making ohmic contact with silicon pads. Conductance of the bridge together with contact pads is enhanced after the stress, although contact pads' surface texture is not modified by the applied pulse.

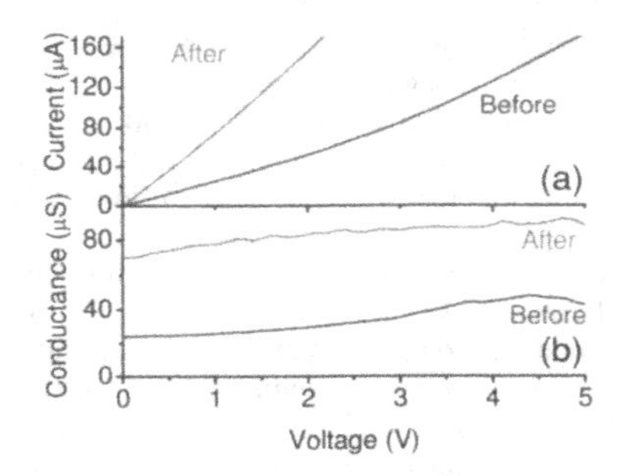

Figure 4. (a) I-V characteristics of the bridge shown in Figure 2b, before and after the electrical stress. (b) Conductance of the bridge extracted from the I-V characteristics.

CONCLUSIONS

A growth from the melt approach utilizing large-amplitude short-duration voltage pulses as a crystallization technique is presented. SEM micrographs and I-t, V-t characteristics during the electrical stress suggest that the bridges are melted, and re-solidified upon termination of the stress. TEM analysis is necessary to verify the crystallinity, crystal orientation and defects.

ACKNOWLEDGMENTS

The authors would like to thank Nathan Henry who fabricated the devices at Cornell Nanofabrication Facility (CNF) as a part of CNF REU program [8]. Imaging facilities of Harvard's Center for Nanoscale Systems (CNS) is used for SEM micrographs.

REFERENCES

[1] R. H. Reuss, B. R. Chalama, A. Moussessian, M. G. Kane, A. Kumar, D. C. Zhang, J. A. Rogers, M. Hatalis, D. Temple, G. Moddel, B. J. Eliasson, M. J. Estes, J. Kunze, E. S. Handy, E. S. Harmon, D. B. Salzman, J. M. Woodall, M. A. Alam, J. Y. Murthy, S. C. Jacobsen, M. Olivier, D. Markus, P. M. Campbell and E. Snow, "Macroelectronics: Perspectives on technology and applications," *Proc IEEE,* vol. 93, pp. 1239–1256, 2005.

[2] S. Wagner, H. Gleskova, I. C. Cheng and M. Wu, "Silicon for thin-film transistors," *Thin Solid Films,* vol. 430, pp. 15-19, 2003.

[3] M. C. McAlpine, R. S. Friedman and C. M. Lieber, "High-Performance Nanowire Electronics and Photonics and Nanoscale Patterning on Flexible Plastic Substrates," *Proceedings of the IEEE, Jul,* vol. 93, pp. 1357-1363, 2005.

[4] T. I. Kamins, *Polycrystalline Silicon for Integrated Circuits and Displays.* Kluwer Academic Publishers, 1998.

[5] C. Boztug, G. Bakan, M. Akbulut, N. Henry, A. Gokirmak and H. Silva, "Numerical modeling of electrothermal effects in silicon nanowires," in Mater, Res, Soc, Symp, Proc, vol. 1083, pp. R04-11, 2008.

[6] The Cornell NanoScale Science & Technology Facility, www.cnf.cornell.edu.

[7] M. Raman, T. Kifle, E. Bhattacharya and K. Bhat, "Physical Model for the Resistivity and Temperature Coefficient of Resistivity in Heavily Doped Polysilicon," *Electron Devices, IEEE Transactions on,* vol. 53, pp. 1885-1892, 2006.

[8] N. Henry, "Crystallization of amorphous silicon nanowires using electromigration and self-heating for TFT applications" *NNIN REU 2006 Research Accomplishments.*

Mater. Res. Soc. Symp. Proc. Vol. 1144 © 2009 Materials Research Society 1144-LL04-02

Three-Dimensional Structure of Helical and Zigzagged Nanowires Using Electron Tomography

Han Sung Kim[1], Yoon Myung[1], Chang Hyun Kim[1], Seung Yong Bae[2], Jae-Pyoung Ahn[3], and Jeunghee Park[1]*

[1] Department of Chemistry, Korea University, Jochiwon 339-700, Korea, Republic of
[2] Chemical Research and Development Center, Samsung Cheil Industry Inc., Uiwang 332-2, Korea, Republic of
[3] Advanced Analysis Center, Korea Institute of Science and Technology, Seoul 136-791, Korea, Republic of

ABSTRACT

Electron tomography and high-resolution transmission electron microscopy were used to characterize the unique three-dimensional structures of helical or zigzagged GaN, $ZnGa_2O_4$ and Zn_2SnO_4 nanowires. The helical GaN nanowires adopt a helical structure that consists of six equivalent <0-111> growth directions with the axial [0001] direction. The $ZnGa_2O_4$ nanosprings have four equivalent <011> growth directions with the [001] axial direction. The zigzagged Zn_2SnO_4 nanowires consisted of linked rhombohedrons structure having the side edges matched to the <011> direction, and the [111] axial direction.

INTRODUCTION

One-dimensional (1D) nanostructures have attracted considerable attention due to their potential use as building blocks for assembling active and integrated nanosystems.[(1)] Recently, interest in helical (or spring-like) and zigzagged nanostructures has been steadily increasing, owing to their attractive morphology and properties. It was demonstrated that helical carbon nanotubes (CNTs) and nanowires (e.g., ZnO, InGaAs/GaAs, Cr) can be used as extremely sensitive mechanical resonators to detect mass and pressure changes.[(2a,2b,5b,17,18)] All of the zigzagged or helical structures in these previous studies were analyzed by examining their two-dimensional (2D) projections using transmission electron microscopy (TEM), which provides a first insight into their size and morphology. There are, however, potentially some cases where important three-dimensional (3D) structural information is missed or erroneous information may be obtained when using simply this technique. Electron tomography, which is a method to reconstruct 3D morphologies from a series of 2D images or projections, has been successfully applied to analyze the morphology of nanoparticles as well as their location in a mesoporous matrix (or carbon nanotubes or nanocomposites).[(21-28)] However, the use of electron tomography to study the 3D geometry of helical or zigzagged nanostructures has not been much reported, despite the ever-increasing research effort devoted to this field. Herein, we report the 3D structures of helical or zigzagged GaN, $ZnGa_2O_4$ and Zn_2SnO_4 nanowires (NWs), obtained by electron tomography and high-resolution TEM. The present work demonstrates their 3D reconstruction images, acquired from a series of 2D projections obtained by high-angle annular dark field (HAADF) scanning TEM (STEM). The results of this study should open up a new field allowing for the investigation of nanostructures with high spatial resolution that could have an impact on our understanding of the growth mechanism and application of nanodevices in many fields in which the crystal structure plays an important role in the final properties.

EXPERIMENT

(1) Materials

The helical or zigzagged GaN, $ZnGa_2O_4$ and Zn_2SnO_4 NWs were synthesized by thermal evaporation method. First, a Ga (99.999%, Aldrich)/GaN (99.99%, Aldrich) mixture was placed in a quartz boat loaded inside a quartz tube reactor. A silicon substrate coated with 1 mM $HAuCl_4 \cdot 3H_2O$ (98%, Aldrich) ethanol solution was positioned at a distance of 10 cm from the quartz boat. The argon gas was allowed to flow at a rate of 500 sccm while raising or lowering the temperature. The temperature of the Ga source was set to 1000 °C, and that of the substrate was approximately 800-850 °C. The NH_3 gas (99.999%) was introduced at a flow rate of 50 sccm for 1 h. Then, the helical GaN NWs were grown. Second, High-purity single-crystalline ZnSe nanowires were synthesized using CoSe (99%, Aldrich)/ZnO (99.98%, Aldrich) on a Au nanoparticles-deposited Si substrate at 800 °C. The pre-grown ZnSe nanowires were placed near the ZnO/Ga (99.99%, Aldrich), and a temperature of 600 or 900 °C was maintained for 10–60 min, thereby producing the $ZnGa_2O_4$ NWs. Finally, a mixture of ZnO (99.999%, Aldrich) and Sn (99.99%, Aldrich) powders was placed inside a quartz tube. The zigzagged Zn_2SnO_4 NWs were grown on the Au-deposited Silicon substrates at 900-1000 °C, by the evaporation of the Zn/Sn source for 1 h under nitrogen gas flow with a rate of 500 sccm.

(2) Characterizations

The morphology, composition, and crystal structure of products were analyzed by scanning electron microscopy (SEM, Hitachi S-4700), field-emission transmission electron microscopy (FEI TECNAI G^2 200 kV), and energy-dispersive X-ray fluorescence spectroscopy (EDX). High-resolution XRD patterns were obtained using the 8C2 beam line of the Pohang Light Source (PLS) with monochromatic radiation (λ=1.54520 Å). XPS (ESCALAB 250, VG Scientifics) using a photon energy of 1486.6 eV (Al Kα) was employed to investigate the electronic states. Synchrotron XPS measurements were also performed at the U7 beam line of the PLS.

(3) Electron Tomography

3D electron tomography was performed using a STEM (FEI Co., Technai F20), with a tilt holder (Dual Orientation Tomography Holder 927, Gatan Co.) and a Fischione model 3000 HAADF detector operated at 200 kV. A series of 110 HAADF-TEM images was collected from +70° to −70° in 1.5° steps under a nominal magnification of 40000~110000×, resulting in a pixel size of 1~3 nm on a computer-controlled sample stage using Xplore3D software (FEI Co.). The images were spatially aligned by a cross-correlation algorithm using Inspect3D software (FEI Co.), and the 3D reconstructions were achieved using a simultaneous iterative reconstruction algorithm (SIRT) from consecutive 2D slices.[29] Visualization was performed using AMIRA 4.0.

RESULTS AND DISCUSSION

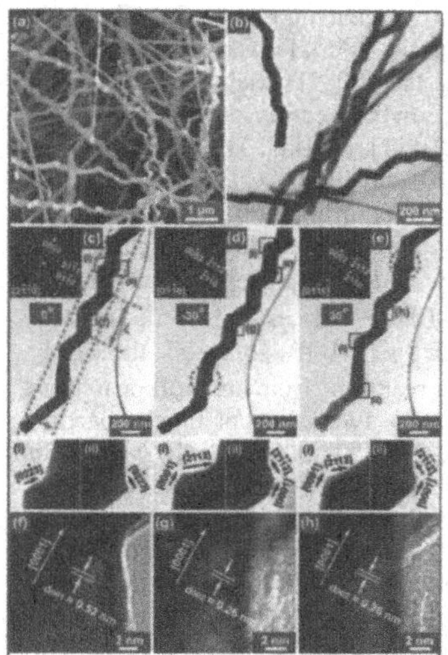

Figure 1. Images of the helical GaN NWs obtained through SEM and TEM

Figure 1a shows the SEM image of the high-density helical GaN NWs synthesized using the vapor transport method. The X-ray diffraction (XRD) and X-ray photoelectron spectroscopy (XPS) data confirm the wurtzite structure GaN NWs, as shown in the Supporting Information, Figure S1.(34) The TEM image reveals all zigzagged structure over the whole NW (Figure 1b). The average diameter (*d*) of the NWs is 100 nm. Figure 1c shows explicitly that the pitch distance (*L* as defined in the figure) is in the wide range of 200-500 nm and the zigzag angle is 120-130°, but the diameter of the helix (*D*) is uniformly 300 nm along the whole NW. We turned the TEM grid holder to rotate the NW around the axial direction. Figure 1d,e corresponds to the TEM images for its -30 and 30° rotations, respectively, showing the maintenance of the zigzagged morphology with the same D value. A series of TEM images for the 10° sequential rotations are shown in the Supporting Information, Figure S2.(34) The corresponding selected-area electron diffraction (SAED) pattern is shown in the insets. The zone axis is [2-1-10] for the 0° turn and [01-10] for the -30 and 30° turns. The [0001] axial direction remains the same for the rotation. This symmetric and periodic right-handed helix structure probably arises from the identical growth direction of the NW blocks (or units) that are directed toward the six equivalent directions of the hexagonal unit cell. The zigzagged direction, resulting from the shared direction of at least two blocks, can be identified using the edge parts ((i) and (ii)), as shown in the lower insets. At the [2-1-10] zone axis, the magnified images show the [02-23] and [0-223] directions. At the [01-10] zone axis, the NW clearly has the [2-1-12], [-2112], and [0001] directions. However, the length of the blocks is very irregular, so that some of the shorter ones appear to be nearly collinear with the others (marked by the circles). This occurs as frequently as 1 per 4 or 5 zigzagged segments and causes an increase in the zigzag angle. Figure 1f-h displays the lattice-resolved images for the parts marked in Figure 1c-e, respectively, proving that this NW is entirely composed of perfect single-crystalline GaN nanocrystals. We observed the same zigzagged directions for other NWs, as shown in the Supporting Information, Figure S3.(34)

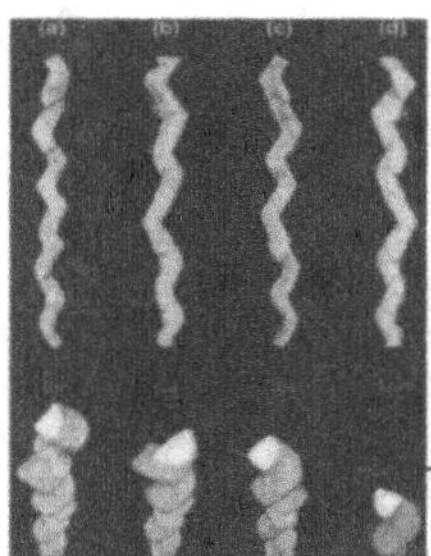

Figure 2 displays the tomographic 3D reconstruction images of the helical GaN NW, as shown in Figure 1c-e, with panels a-d being the images for the 90° sequential turns. The corresponding movie is supplied in Supporting Information, Movie S1.(34) The sliced views along the NW (as marked in Figure 2a, i-iv), reveal the triangular cross-section. Supporting Information, Movie S2 provides the vertical-direction rotational view.(34) The additional tomography data are supplied in the Supporting Information, Figure S4 and Movies S3 and S4.(34)

Figure 2. Images of the helical GaN NW obtained through tomographic 3D reconstruction.

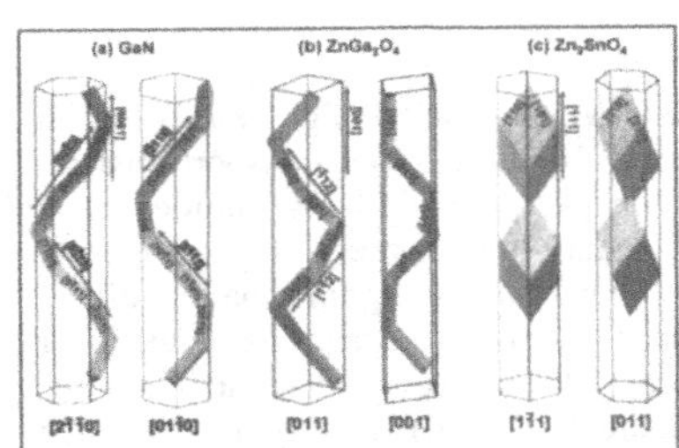
Figure 3. Schematic model constructed for the helical GaN and $ZnGa_2O_4$ NWs and zigzagged Zn_2SnO_4 NWs.

On the basis of the assigned direction, we build a schematic model as shown in Figure 3a. The NW blocks stack along the [0001] axial direction. The blocks stack along the [0001] axial direction. They have six <0-111> growth directions: [-1101], [0-111], [1-101], [10-11], [01-11], [-1011]. The angle between <0-111> and [01-10] directions is 43.2°, as obtained from stereogram (Supporting Information, Figure S5).[34] The triangular cross-section suggests that these NW blocks would be enclosed with ±(01-11), ±(1-101), and ±(-1011) surfaces. When the incident beam is projected at the [2-1-10] zone axis, the TEM image shows two [02-23] and [0-223] zigzagged directions, resulting from the collinear three adjacent blocks (left image). As the NW is rotated by 30°, the two adjacent blocks become collinear and share the same growth direction, [2-1-12] and [-2112], at the [01-10] zone axis (right image). At this zone axis, the two blocks line up along the [0001] axial direction. Zigzagged GaN and coiled GaInN NWs were reported by other research groups, but the 3D structure of the helical GaN NWs has not previously been analyzed.[9,11] For the helical ZnO NWs, Wang and co-workers suggested a model in which the growth is led by the Zn-terminated (0001) front surface and the six equivalent <0-111> growth directions are formed to reduce the electrostatic interaction energy caused by the ±{01-11} polar surfaces of the NWs.[5a] This model may be adopted for the present helical GaN NWs having the Ga (or N)-terminated ±(1-101) and ±(-1011) polar side surfaces that attracted each other.

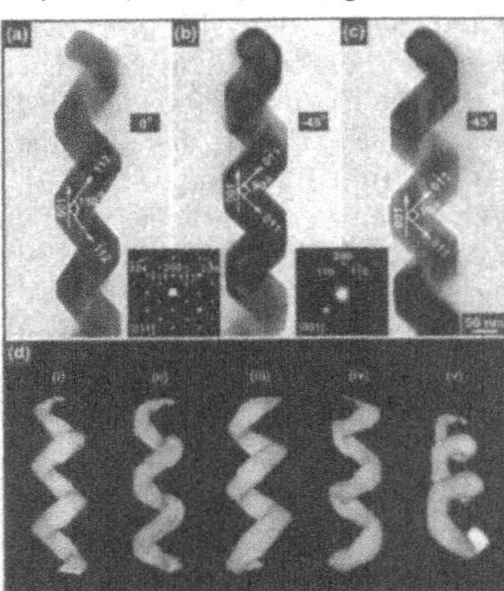
Figure 4. Images of the helical $ZnGa_2O_4$ NWs obtained through SEM and TEM

We reported spinel-structure $ZnGa_2O_4$ nanosprings (right-handed), synthesized using ZnSe NWs as templates.[10] They have explicitly the [1-12]/[-112] zigzagged directions (with a zigzag angle of 110°) at the [011] zone axis (Figure 4a and inset). They have a more uniform pitch length than the GaN NWs. When the nanospring is rotated by -45 or 45°, the [0-11]/[011] zigzagged directions (with a zigzag angle of 90°) are observed at the [001] zone axis (Figure 4b,c and insets). The [001] axial direction maintains along the whole length. A series of TEM images for the 10° rotation are shown in the Supporting Information, Figure S6.[34] Figure 4d corresponds to the tomographic 3D reconstruction images, panels i-iv are the images for sequential 45° rotation, and v is for its top view, showing their square cross-section. The corresponding movies are displayed in Supporting Information, Movies S5 and S6.[34] The schematic model is shown in Figure 3b. All these results confirm our previous structural assignment.[10]

The SEM image of Zn_2SnO_4 NWs shows their diamond chain-like morphology (Figure 5a). The TEM image reveals that all of the NWs consist of linked rhombohedral nanocrystals (Figure 5b). The EDX, XRD, and XPS data confirm a face-centered cubic spinel structure, as shown in the Supporting Information, Figures S7 and S8.[34] For a selected NW, the zigzag angle is 125° and

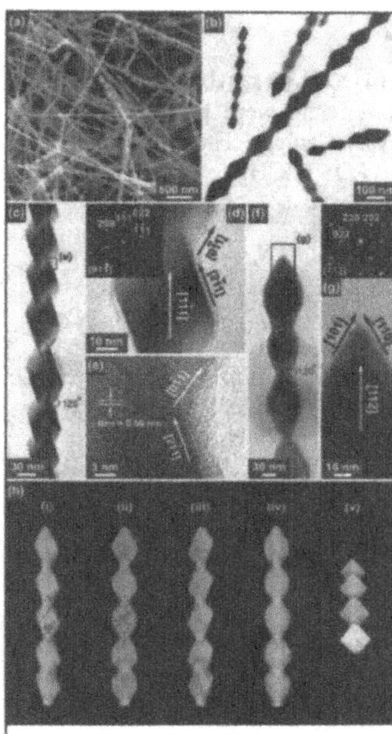

Figure 4. Images of the helical Zn_2SnO_4 NWs obtained through SEM and TEM

the edge angle of the rhombohedral nanocrystals is 55°, as marked by the red-color lines (Figure 5c). The TEM image and SAED pattern (at the [011] zone axis) reveal the [0-11] and [2-11] zigzagged direction and the [1-11] wire axis (Figure 5d and inset). The lattice-resolved image confirms their single-crystalline nature (Figure 5e). Another NW was found to have a zigzag angle of 120° (Figure 5f). Its tip has an angle of 60° (Figure 5g). The SAED pattern at the [-111] zone axis indicates the [101] and [110] zigzagged directions and the [211] axial direction (inset). We measured a series of TEM images for the sequential rotation of another zigzagged NW (Supporting Information, Figure S9).(34) As the NW is rotated by 30°, the zone axis changes from [011] to [-111], and the zigzag angle changes from 125 to 120°. Therefore, all of the NWs, we observed have the same zigzagged structure. Figure 5h displays the tomographic 3D reconstruction images, and the corresponding movie is supplied in Supporting Information, Movies S7 and S8.(34) Images i-iv correspond to sequential turns, showing the change of rhombohedral angle from 55 to 60°. The top view (v) reveals its rhombic-shaped cross-section. A schematic model is shown in Figure 3c. We suggest that the six apexes of the rhombohedral blocks are directed toward the six equivalent <110> directions of the hexagonal units and the two apexes along the [111] axial direction. At the [-111] zone axis, the TEM image shows the [110]/[101] zigzagged directions and the [211] axial direction, which overlaps with the [111] direction (left image). As the NW is rotated by 30°, the zone axis becomes [011] (right image). Then, one side edge coincides with the [211] direction due to the collinear [110]/[101] directions. It is noteworthy that the assignment of the axial direction for such zigzagged NWs can often be incorrect when only the 2D projection image is used. Zigzagged Zn_2SnO_4 NWs were previously reported, commonly showing the [111] axial direction.(30-33) The zigzagged direction was observed to be [211] or [011], which is consistent with the present NWs.

CONCLUSIONS

In summary, tomographic reconstruction and HRTEM images were used to characterize the unique 3D structures of right-handed helical GaN NWs that consist of six equivalent <0111> growth directions with the [0001] axial direction. We also present the tomography data for zigzagged CdO and Zn_2SnO_4 NWs consisting of rhombohedral shaped nanocrystals in which the side edges match the six equivalent <110> directions and the axial direction is [111] were also studied. $ZnGa_2O_4$ nanosprings, having four equivalent <011> growth directions with the axial [001] direction. Therefore, we suggest that tomographic 3D construction allows for precise structure analysis, particularly for helical/zigzagged 1D nanostructures.

ACKNOWLEDGMENTS

This work was supported by the KRF (Project No. R14-2004-033-01003-0; R02-2004-000-100250; 2004-015-C00125). The SEM and TEM measurements were performed at the Korea Institute of Science and Technology (Advanced Analysis Center). The experiments of XRD and XPS at the PLS were supported in part by MOST and POSTECH.

REFERENCES

(1) (a) Hu, J.; Odom, T. W.; Lieber, C. M. *Acc. Chem. Res.* **1999**, *32*, 435. (b) Gudiksen, M. S.; Lauhon, L. J.; Wang, J.; Smith, D. C.; Lieber, C. M. *Nature (London)* **2002**, *415*, 617. (c) Duan, X.; Huang, Y.; Agarwal, R.; Lieber, C. M. *Nature (London)* **2003**, *421*, 241.
(2) (a) Poggi, M. A.; Boyles, J. S.; Bottomley, L. A.; McFarland, A. W.; Colton, J. S.; Nguyen, C. V.; Stevens, R. M.; Lillehei, P. T. *Nano Lett.* **2004**, *4*, 1009. (b) Volodin, A.; Buntinx, D.; Ahlskog, M.; Fonseca, A.; Nagy, J. B.; Van Haesendonck, C. *Nano Lett.* **2004**, *4*, 1775.
(3) Tang, Y. H.; Zhang, Y. F.; Wang, N.; Lee, C. S.; Han, X. D.; Bello, I.; Lee, S. T. *J. Appl. Phys.* **1999**, *85*, 7981.
(4) Zhang, H.-F.; Wang, C.-M.; Wang, L.-S. *Nano Lett.* **2002**, *2*, 941.
(5) (a) Yang, R.; Ding, Y.; Wang, Z. L. *Nano Lett.* **2004**, *4*, 1309. (b) Gao, P. X.; Mai, W.; Wang, Z. L. *Nano Lett.* **2006**, *6*, 2536.
(6) Zhan, J.; Bando, Y.; Hu, J.; Xu, F.; Golberg, D. *Small.* **2005**, *1*, 883.
(7) Duan, J.; Yang, S.; Liu, H.; Gong, J.; Huang, H.; Zhao, X.; Zhang, R.; Du, Y. *J. Am. Chem. Soc.* **2005**, *127*, 6180.
(8) Duan, J. H.; Yang, S. G.; Liu, H. W.; Gong, J. F.; Huang, H. B.; Zhao, X. N.; Zhang, R.; Du, Y. W. *J. Phys. Chem. B* **2005**, *109*, 3701.
(9) Zhou, X. T.; Sham, T. K.; Shan, Y. Y.; Duan, X. F.; Lee, S. T.; Rogenberg, R. A. *J. Appl. Phys.* **2005**, *97*, 104315.
(10) Bae, S. Y.; Lee, J.; Jung, H. S.; Park, J.; Ahn, J. -P. *J. Am. Chem. Soc.* **2005**, *127*, 10802.
(11) Cai, X. M.; Leung, Y. H.; Cheung, K. Y.; Tam, K. H.; Djuris˘ic´, A. B.; Xie, M. H.; Chen, H. Y.; Gwo, S. *Nanotechnology* **2006**, *17*, 2330.
(12) Moore, D.; Ding, Y.; Wang, Z. L. *Angew. Chem., Int. Ed.* **2006**, *45*, 5150.
(13) Zhang, L.; Ruh, E.; Grűtzmacher, D.; Dong, L.; Bell, D. J.; Nelson, B. J.; Schönenberger, C. *Nano Lett.* **2006**, *6*, 1311.
(14) Yu, D.; Wu, J.; Gu, Q.; Park, H. *J. Am. Chem. Soc.* **2006**, *128*, 8148.
(15) He, Y.; Fu, J.; Zhang, Y.; Zhao, Y.; Zhang, L.; Xia, A.; Cai, J. *Small* **2007**, *3*, 153.
(16) Shen, G. Z.; Bando, Y.; Zhi, C. Y.; Yuan, X. L.; Sekiguchi, T.; Golberg, D. *Appl. Phys. Lett.* **2006**, *88*, 243106.
(17) Bell, D. J.; Dong, L.; Nelson, B. J.; Golling, M.; Zhang, L.; Grűtzmacher, D. *Nano Lett.* **2006**, *6*, 725.
(18) Kesapragada, S. V.; Victor, P.; Nalamasu, O.; Gall, D. *Nano Lett.* **2006**, *6*, 854.
(19) Zhai, T.; Gu, Z.; Yang, W.; Zhang, X.; Huang, J.; Zhao, Y.; Yu, D.; Fu, H.; Ma, Y.; Yao, J. *Nanotechnology* **2006**, *17*, 4644.
(20) Nath, M.; Parkinson, B. A. *J. Am. Chem. Soc.* **2007**, *129*, 11302.
(21) Möbus, G.; Doole, R. C.; Inkson, B. J. *Ultramicroscopy* **2003**, *96*, 433.
(22) Thomas, J. M.; Midgley, P. A. *Chem. Comm.* **2004**, 1253.
(23) Koster, A. J.; Ziese, U.; Verkleij, A. J.; Janssen, A. H.; de Jong, K. P. *J. Phys. Chem. B* **2000**, *104*, 9368.

(24) Kanaras, A. G.; Sönnichsen, C.; Liu, H.; Alivisatos, A. P. *Nano Lett.* **2005**, *5*, 2164.
(25) Bae, A.-H.; Numata, M.; Hasegawa, T.; Li, C.; Kaneko, K.; Sakurai, K.; Shinkai, S. *Angew. Chem., Int. Ed.* **2005**, *44*. 2030.
(26) Yoshizawa, N.; Tanaike, O.; Hatori, H.; Yoshikawa, K.; Kondo, A.; Abe, T. *Carbon* **2006**, *44*, 2558.
(27) Ersen, O.; Werckmann, J.; Houlle, M.; Ledoux, M.-J.; Pham-Huu, C. *Nano Lett.* **2007**, *7*, 1898.
(28) Park, J.-B.; Lee, J. H.; Choi, H.-R. *Appl. Phys. Lett.* **2007**, *90*, 093111.
(29) Guckenberger, R. *Ultramicroscopy* **1982**, *9*, 167.
(30) Wang, J. X.; Xie, S. S.; Gao, Y.; Yan, X. Q.; Liu, D. F.; Yuan, H. J.; Zhou, Z. P.; Song, L.; Liu, L. F.; Zhou, W. Y.; Wang, G. *J. Cryst. Growth* **2004**, *267*, 177.
(31) Jie, J.; Wang, G.; Han, X.; Fang, J.; Yu, Q.; Liao, Y.; Xu, B.; Wang, Q.; Hou, J. G. *J. Phys. Chem. B* **2004**, *108*, 8249.
(32) Li, Y.; Ma, X. L. *Phys. Status Solidi A* **2005**, *202*, 435.
(33) Jeedigunta, S.; Singh, M. K.; Kumar, A.; Shamsuzzoha, M. *J. Nanosci. Nanotechnol.* **2007**, *7*, 486.
(34) Kim, H. S.; Hwang, S. O.; Myung, Y.; Park, J.; Bae, S. Y.; Ahn, J.-P. *Nano Lett.* **2008**, *8*, 551.

Mater. Res. Soc. Symp. Proc. Vol. 1144 © 2009 Materials Research Society 1144-LL04-11

Anodization of NbN

Travis L. Wade, Damien Lucot, Abdullah Al Ahmari, Mihaela-Cristina Ciornei, Jean-Eric Wegrowe, and Kees van der Beek
Laboratoires des Solides Irradiés, ECOLE Polytechnique, 91128 Palaiseau, France

ABSTRACT

This brief report discusses anodization of NbN using a recipe developed for the anodization of Nb. The anodization slowly oxidizes the NbN until the NbN wire diameters are small enough that their conductance is no longer ohmic. This indicates that capacitive defects are dominating the electronic conduction. Ideally, this could be a simple approach to form weak-link Josephson Junctions. Preliminary room temperature IV curves and R(T) measurements of this new system are shown.

INTRODUCTION

The dimensions of superconducting materials are expected to have an effect on their properties.[1] Niobium nitride (NbN) is a superconducting material with a Tc of 13 K. It is used in SQUIDs and for single photon detectors.[2] NbN lines of 70 nm are formed on SiO_2 passivated silicon by e-beam lithography. These dimensions are the limit of our facilities. Current-voltage measurements of the lines are Ohmic. NbN films are anodized normal to their plane and lines are anodized perpendicular to their axis using the same solution as was developed for the anodization of Nb.[3] The anodization of the Nb resulted in a network of nanopores similar to the results of the anodization of Al. The anodization slowly oxidizes the NbN until the NbN wire diameters are small enough that their room temperature conductance has increased. This indicates that capacitive defects are dominating the electronic conduction. This could be a simple approach to form weak-link Josephson Junctions without the use of exotic lithography or scanning probe techniques such as atomic force microscopy (AFM).[2,4]

EXPERIMENTAL

NbN bulk films on SiO_2/Si and 70 nm diameter lines on SiO_2/Si were mounted on plexiglass supports and contacted with aluminum wires using silver paste. The contacts were covered with paraffin to protect them from the anodization. The anodization solution was 1% HF in 1 M H_3PO_4 and the sample polarization was +5 V. The solution was constantly stirred by a magnetic stir bar. The electron microscope photos were made with a Hitachi field-emission scanning electron microscope, FESEM. The 2-point current/voltage (IV) and resistance versus temperature R(T) measurements were made with a Keithley Instruments 2400 multimeter and the lower temperature measurements using a liquid He cryostat with a Lake Shore temperature controller and Keithley 2182 nanovoltmeter.

RESULTS

Figure 1 shows the sample resistance versus time for a 70 nm thick NbN film during anodization. It required 3 hours and 20 minutes to anodize the entire 70 nm film. This corresponds to an anodization rate of about 20 nm/h.

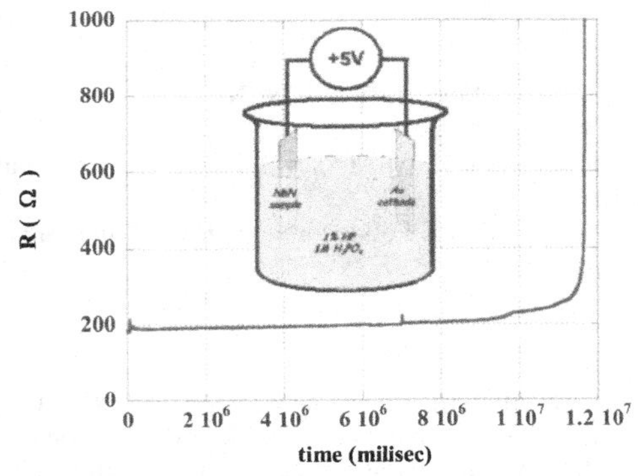

FIG. 1. Resistance versus anodization time for a NbN 70 nm thick film. The inset drawing shows the simplified experimental configuration.

Figure 2 shows the anodization results of the 70 nm NbN film. Area A was not anodized and area B anodized. The anodization does not produce an ordered nanoporous oxide as is the case for the anodization of Nb metal.[2] There is clearly a change in morphology and it appears that the thickness of the anodized area has increased. The non-anodized area (figure 2A) of the NbN film is composed of <10 nm bumps, whereas, the anodized area (figure 2B) shows a larger grain size with cracks between the grains.

FIG. 2. FESEM photo of the interface of NbN, A, and the anodized area, B.

Figure 3 shows a 70 nm NbN line that was anodized for three hours at +5 V. The line is broken at one point and it is much larger then the original 70 nm width; more like 200 nm.

The room temperature resistance of the line was 170 kΩ. The low temperature resistance exponentially increased to several MΩ.

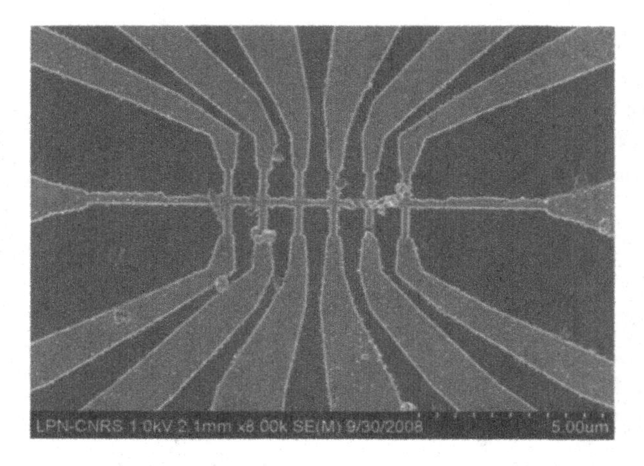

FIG. 3. SEM photo of a NbN line anodized at +5 V for 3 hours.

Figure 4 is the room temperature IV plots for an anodized NbN line like that shown in Figure 2. The sample was anodized for one hour at +5 V. The initial resistance was 7 K ohms and final resistance was 18 kΩ. This corresponds to a diameter change of 70 to 50 nm as is also seen in the temperature measurements, Figure 6 and 7.

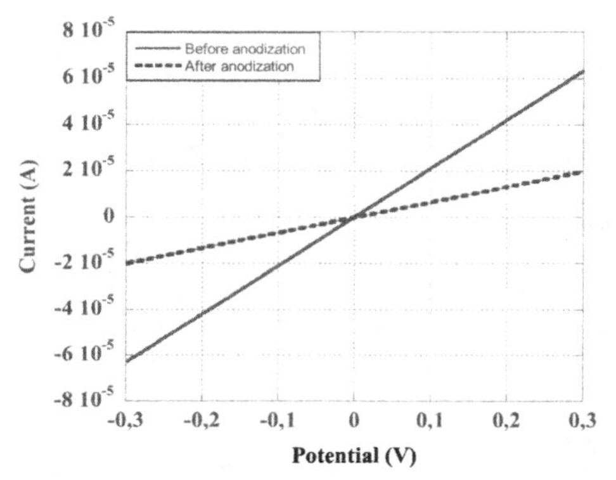

FIG. 4. Room temperature IV plots of a 10 micron long 70 nm width NbN line anodized for 1 hour at +5 V.

Figure 5 is a plot of the resistance versus temperature for a 70 nm thick film. There is a very clear superconducting transition temperature (T_c), depending on the measuring current, starting 11 K. For 0.1 mA the measured T_c is 11 K; the transition moves down to 10 K at 16 mA measuring current. At the base of the transition for the high current curves there is evidence of quasi-particle generation i.e. depairing.

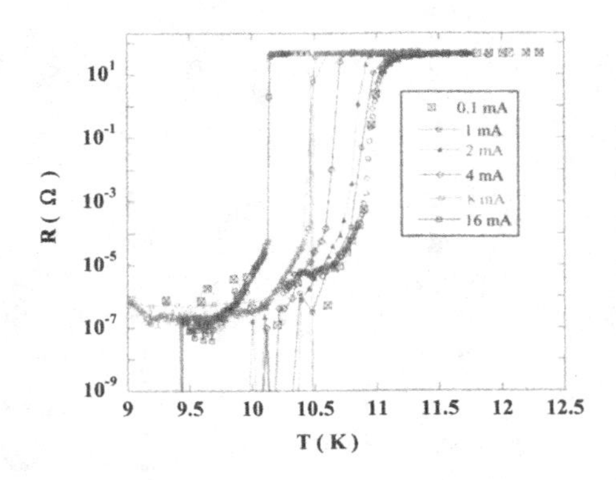

FIG. 5. R(T) plot as a function of current for a 70 nm thick NbN film.

Figure 6 shows the R(T) measurements for the anodized NbN line from figure 4. There is evidence of at least five phase-slip centres that go to normal at lower temperatures with increasing current. At the base of the transition for the high current curves there is evidence of quasi-particle generation i.e. depairing as was seen in the film. Unlike the film, however, T_c exceeds 12 K and the transition does not move to lower temperatures with higher measuring currents.

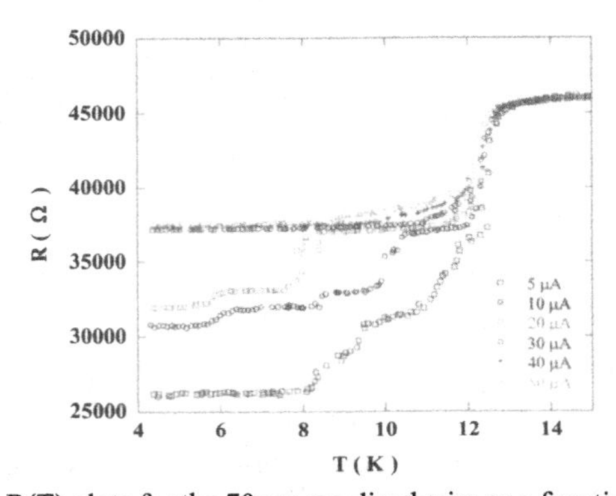

FIG. 6. R(T) plots for the 70 nm anodized wire as a function of current.

Figure 7 is a combined R(T) plot for the film, figure 5, and the line, figure 6. The blue curve is the film, the red curve the anodized nanowire, and the orange curve the contact resistance of the anodized wire. This was extracted by subtracting that part of the line resistance attributable to the remaining NbN core. The resistive fraction due to the remaining NbN is determined from the extrapolation to zero current of the cumulative change of the resistance between 4 K and T_c (as depicted in Fig. 6), assuming that for $I \rightarrow 0$, a zero resistance state is ultimately reached in the NbN. The resistance of the film is plotted on the right and the line on the left. The resistance of the NbN wire in the normal state has the same behavior as that of the bulk; only the cross-section is narrower, such that the resistance is increased by a factor of 420. The conduction is, therefore, still three-dimensional. The remaining resistance, attributable to the contacts, is exponential both in temperature and current (inserted plot) which is consistent with a tunnel barrier.

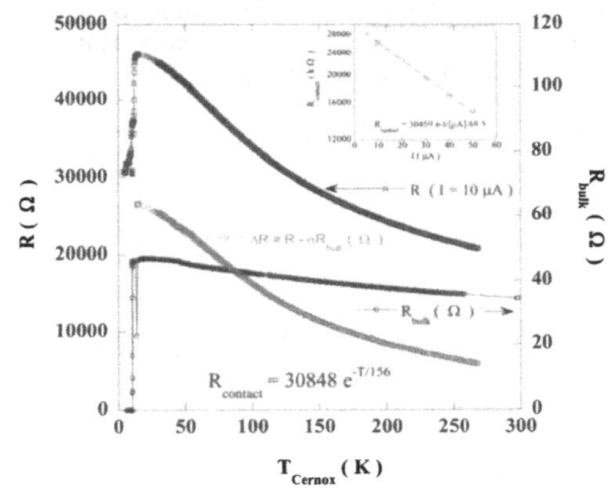

FIG. 7. Resistance versus temperature plots for a 70 nm thick thin film and a 70 nm diameter anodized nanowire.

CONCLUSIONS

Anodization of NbN at +5 volts in 1% HF in 1 M H_3PO_4 results in the decrease in thickness (diameter) of the NbN at a rate of 20 nm/h. The structure of the anodized area is not an ordered nanoporous network as is the case for the anodization of Nb in the same solution at +2.5V.1 Anodization of the NbN at +2.5 V had no effect after several hours.

Anodization of the 70 nm diameter NbN wire for one hour reduced its diameter by 20 nm, 70 nm to 50 nm. The resistance of the wire increased from 7 kΩ to 18 kΩ. We subtract the resistance of the remaining NbN core from that of the wire as a whole to reveal a contact resistance of 3 kΩ. This contact resistance is exponential in current and temperature which is consistent with a tunnel barrier.

REFERENCES

1) Current-voltage characteristics of quasi-one-dimensional superconductors: an S-shaped curve in the constant voltage regime, Vodolazov, D. Y.; Peeters, F. M.; Piraux, L.; Matefi-Tempfli, S.; Michotte, S., PHYSICAL REVIEW LETTERS Vol. 91, p.15001, 2003.
2) Niobium and niobium nitride SQUIDs based anodized nanobridges made with an atomic force microscope, Faucher, M; Fournier, T; Pannetier, B; Thirion, C; Wernsdorfer, W; Villegier, J. C.; Bouchiat, V., Conference on Physics and Applications of Superconducting Quantum Interference (SQUID 2001), AUG 31-SEP 03, 2001 Stenungsbaden Sweden, PHYSICA C-SUPERCONDUCTIVITY AND ITS APPLICATIONS Vol. 368(1-4) p. 211-217, 2002.
3) Nanoporous niobium oxide for label-free detection of DNA hybridization events, Jinsub, C.; Lim, J. H.; Rho, S.; Jahng, D.; Lee, J.; Kim, K. J., TALANTA, Vol. 74 p. 1056-1059, 2008
4) Multicontact measurements of a superconducting Sn nanowire, Lucot, D.; Pierre, F.; Mailly, D.; Yu-Zhang, K.; Michotte, S.; F. de Menten de Horne; Piraux, L., APPLIED PHYSICS LETTERS Vol. 91(4) p. 042502 JUL 2007.

Mater. Res. Soc. Symp. Proc. Vol. 1144 © 2009 Materials Research Society 1144-LL04-15

Langmuir-Blodgett Films of One-Dimensional Nanowires Composed of Amphiphilic Tetrathiafulvalenes and Electron Acceptor

Yoko TATEWAKI[1], Junko TAKIZAWA[1], Tatsuya HATANAKA[2], Mutsumi KIMURA[1,2] and Hirofusa SHIRAI[1]
[1]Collaborative Innovation Center for Nanotech FIBER, Shinshu University, Ueda 386-8567, Japan
[2]Department of Functional Polymer Science, Faculty of Textile Science and Technology, Shinshu University, Ueda 386-8567, Japan

ABSTRACT

We found that novel amphiphilic tetrathiafulvalene (TTF; **1**) organized into supramolecular assemblies by the addition of electron acceptor 2,3,5,6-tetrafluoro-7,7',8,8'-tetracyano-*p*-quinodimethane (F_4TCNQ) at the air-water interface. The assemblies on the air-water interface can be transferred onto the surface of mica and the morphologies of transferred films were investigated by AFM. The morphology of Langmuir-Blodgett films strongly depended on the ratio between **1** and F_4TCNQ. Film of 1 : 1 mixture of **1** and F_4TCNQ formed nanoscopic wires having an average dimension of 5.0 (height) x 70 (width) x 3000 (length) nm. On the other hand, the film of 2 : 1 mixture of **1** and F_4TCNQ showed the mixed domains of wires and flat monolayers. Thus, the completely charge-transferred complex ($\mathbf{1}^+$)(F_4TCNQ^-) formed one-dimensional nanowires having a micrometer length. Room temperature electrical conductivities of cast film for 1 : 1 and 2 : 1 mixed layers of **1** and F_4TCNQ were 2.4×10^{-4} and 8.0×10^{-5} S cm^{-1}, respectively.

INTRODUCTION

One dimensional nano-strucrures like nanowire and nanotube are one of important building-components in nanodevices [1,2]. Molecular-assembly nanomaterials are constructed utilizing intermolecular interactions such as charge-transfer (CT), hydrogen-bonding and coordinate bonding interactions [3-6]. The Langmuir-Blodgett (LB) technique is promising for building a conduction path. This technique provides a way of arranging the molecules in a fixed configuration, which can enhance the occurrence of intermolecular interactions. These characteristics of the LB method can give rise to well ordered monolayer packing in a two dimensional layer [7]. Here we reported the LB films composed of different mixing ratio. The morphologies in mixing ratio between **1** and F_4TCNQ from 1 : 1 to 2 : 1 are expected to change the shape of surface domains and electronic state[8].

EXPERIMENT

The structures of 1 and F_4TCNQ were shown in Scheme 1. 1 was prepared by the procedure described elsewhere [9,10]. F_4TCNQ was purchased from Tokyo Chemical Industry Co., LTD. The Langmuir films at the air-water interface and the LB films were prepared by using a conventional LB trough (USI, USI-3-22) equipped with two symmetrical moving barriers. Pure water was used as a subphase. The charge transfer complex solution of 1 and

F_4TCNQ was prepared by mixing of 1 in $CHCl_3$ (1 mM) with F_4TCNQ in CH_3CN (1 mM). This solution was spread on the pure water. The floating monolayers were remaining for 20 min after spreading of solution. Surface pressure-area per molecule (π - A) isothermes were recorded at 291 K. AFM images were taken using a JSPM-5400 (JEOL) operating at a dynamic force mode. The films of 1 : 1 mixture and 2 : 1 mixture of 1 and F_4TCNQ were transferred onto a freshly cleaved mica surface at a surface pressure at 20 mN m^{-1}, only TTF was transferred at 7 mN m^{-1} by a single up-stroke withdrawal. UV-Vis spectra were recorded on a Hitachi U-4100. Hydrophobic treatments of glass and quartz substrates were made by evaporation of hexamethyldisilazane (HMDS). The films of HMDS were deposited onto quartz substrates (30 x 10 mm^2) for the UV-Vis measurements. The current-voltage curves of cast films were measured using the DC two-probe method. Gold electrodes with a 500 μm gap were deposited by the vacuum evaporation on the glass substrates.

Scheme 1 Molecular structures of **1** and F_4TCNQ.

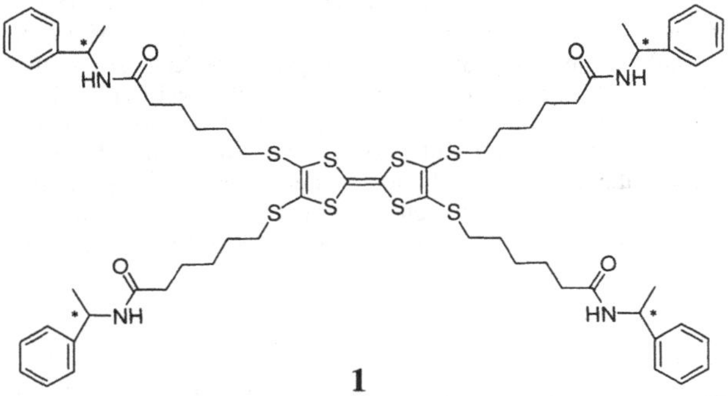

HN NH S S S S S S S S O O O O NH HN

1

NC CN NC CN F F F F

F_4TCNQ

DISCUSSION

Figure 1 shows π - A isotherms of the Langmuir film of 1 : 1 mixture of **1** and F_4TCNQ and neutral donor **1** on subphase of pure water (T = 291 K). Formations of stable floating monolayers were confirmed by sharp increases of the surface pressure during compression. The surface area extrapolated at 0 mN m^{-1} of the 1 : 1 mixed the monolayer on pure water was $A_0 =$ 1.08 nm^2, a kink was observed at around 25 mNm^{-1}. The π - A isotherm of the Langmuir film of 2 : 1 mixture of **1** and F_4TCNQ shows same behavior with that of 1 : 1 mixture. But the film of neutral **1** did not the hard film because no surface pressure arise over 15 mNm^{-1}, the pressure of kink appeared at 15 mNm^{-1}. The Langmuir films of 1 : 1 mixture of **1** and F_4TCNQ and neutral **1** were transferred at 20 or 7 mNm^{-1}. The area per molecule of 1 : 1 mixture of **1** and F_4TCNQ at $\pi = 20$ mNm^{-1} (A_{20}) was 0.36 nm^2, and it of neutral **1** at 7 mNm^{-1} (A_7) was 0.76 nm^2, respectively.

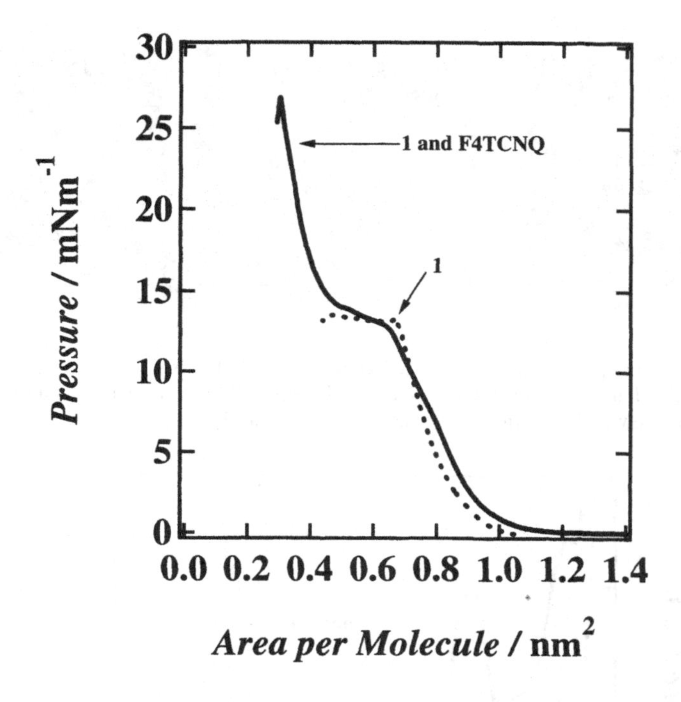

Fig. 1. π-*A* isotherms of 1 : 1 mixed monolayers of **1** and F_4TCNQ and neutral donor **1** on pure water at 291K.

Figure 2 shows AFM images of the film of 1 : 1 and 2 : 1 mixture of **1** and F_4TCNQ transferred onto freshly cleaved mica by a single withdrawal at a fixed surface pressure of 20 mNm^{-1}, the film of the neutral donor **1** was transferred at 7 mNm^{-1}. The completely ionized ($\mathbf{1}^+$)(F_4TCNQ^-) formed oriented nanowires with a typical dimension of 5.0 (height) x 70 (width) x 3000 (length) nm^3 on mica (Fig. 2a). The AFM images of the film of 2 : 1 mixture of **1** and F_4TCNQ and neutral **1** show the 2D film (Fig. 2b) and the circle domain structure of 2D nanowire bundles with a typical dimension of ~300 x 300 nm^2 (Fig. 2c). There are two possibilities on the electronic state of nanostructures; the formation of $(\mathbf{1}^+)_2(F_4TCNQ^-)$ having partially charge-transferred state or charge separated state, and the phase separation between ($\mathbf{1}^+$)(F_4TCNQ^-) and neutral **1**. Phase separation of 2 : 1 mixed LB film was supported by the AFM images and spectroscopic measurement.

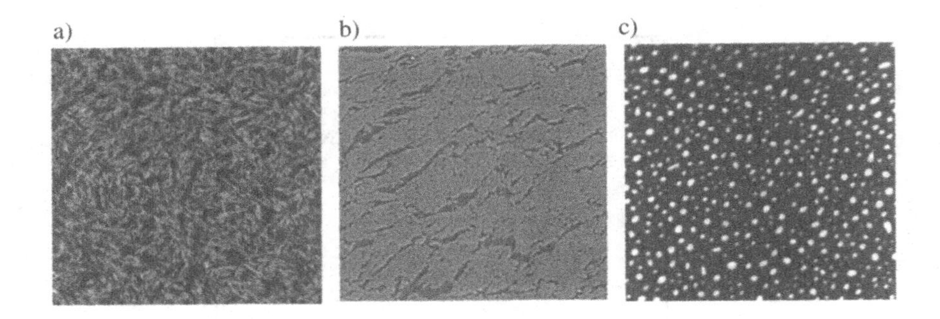

Fig. 2. Surface morphologies of transferred films of a) 1 : 1 mixed monolayers of **1** and F_4TCNQ, b) 2 : 1 mixed monolayers of **1** and F_4TCNQ at 20 mNm^{-1} and c) neutral **1** at 7 mNm^{-1} by a single withdrawal from pure water. The scales are 10 x 10 μm^2.

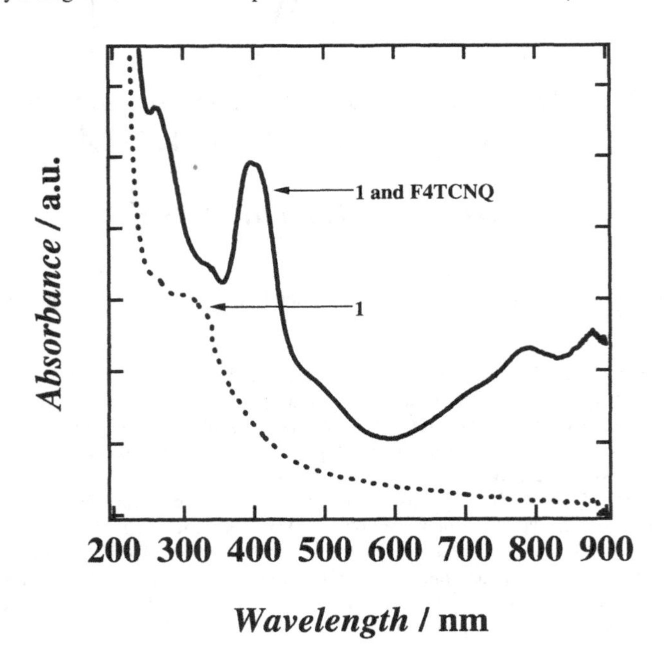

Fig. 3. UV-Vis spectra of 20-layers LB films of 1:1 mixture of **1** and F_4TCNQ and neutral donor **1** transferred from pure water.

UV-Vis spectrum of 20-layers LB films of 1 : 1 mixture of **1** and F_4TCNQ and neutral **1** were measured in the energy region from 200 to 900 (Fig. 3). Two sharp bands at 777 and 882 nm are assigned to the intramolecular transition of F_4TCNQ^- anion radical, which were observed at 763 and 877 cm^{-1} in CH_3CN solution. The bands appeared at 496 and 415 (396) nm were intramolecular transitions of cation radical of 1^+ and anion radical of F_4TCNQ^- [11, 12]. Broad absorption around 900 nm was overlapped with two sharp bands, which was assigned to the intramolecular CT transition of ionized **1** derivative. When donor **1** forms partially charge transferred state of 1^+, the CT transition should appear at over 900 nm [13 - 16]. The absorption band at 336 nm, which is exhibited in neutral **1** LB film, is observed in 2 : 1 mixed LB film. Therefore, 2 : 1 mixed LB films were composed of $(1^+)(F_4TCNQ^-)$ and neutral **1**.

These I-V curves of the cast films of 1 : 1 and 2 : 1 mixture of **1** and F_4TCNQ were measured. The electrical conductivities at room temperature (σ_{RT}) for the cast films of 1 : 1 and 2:1 were 2.4×10^{-4} and 8.0×10^{-5} Scm^{-1}, respectively. The presence of neutral donor **1** in film should interrupt the conduction paths of CT complex, resulting in the lower conductivity.

CONCLUSIONS

Langmuir-Blodgett (LB) films of 1 : 1 mixing ratio between an electron donor of **1** and an electron acceptor of F_4TCNQ were formed on substrate surface. The surface morphologies of the LB films of 1 : 1 and 2 : 1 mixture of **1** and F_4TCNQ show the molecular-assembly nanowires and the 2D film composed of the nanowire of (**1**)(F_4TCNQ) and circle domains of neutral **1**. The LB films of neutral **1** show the nanodot. The electronic spectrum clearly showed that the nanowire composed of 1 : 1 mixture of **1** and F_4TCNQ was CT complex. Electrical conductivity of the cast film of 2 : 1 mixture of **1** and F_4TCNQ was small compared to 1 : 1 mixed film. The neutral **1** species suppressed the formation of conduction paths of $(\mathbf{1})_2(F_4TCNQ)$ in the LB film.

ACKNOWLEDGMENTS

This work was supported by a project for "Creation of Innovation Centers for Advanced Interdisciplinary Research Areas (Shinshu University)" in Special Coordination Funds for Promoting Science and Technology from the Ministries of Education, Culture, Sports, Science and Technology of Japan, the Mazda foundation, and the Yamamura fellowship.

REFERENCES

1. Nanoscale Materials in Chemistry, edited by K. L. Klabunde (Wiley & Sons, NY, 2001).
2. I. W. Hamley, Introduction to Soft Matter - Polymers, Colloids, Amphiphiles and Liquid Crystals, John Wiley & Sons, Weinheim (2000).
3. Ph. Leclere, M Surin, P. Vivlle, R. Lazzaroni, A.F.M. Kilbringer, O. Hanze, W. J. Feast, M. Cavallini, F. Biscarini, A.P.H.J. Schenning and E.W. Meijer, Chem. Mater. **16**, 4452 (2004).
4. K.Akagi, G. Piao, S. Kaneko, K. Sakamaki, H. Shirakawa and M. Kyotani, Science **282**, 1683 (1998).
5. F.J.M. Hoeben, P. Jonkheijm, E.W. Meijer and A.P.H.J. Schenning, Chem. Rev. **105**, 1491 (2005).
6. Y. Yan, Z Yu, Y. Huang, W. Yuan and Z. Wei, Adv. Mater. **19**, 3353 (2007).

7. M. R. Bryce and M. C. Petty, Nature **374**, 771 (1995).
8. The F_4TCNQ shows a reduction potential at $E_{1/2}(1) = 0.70$ V *vs.* SCE in 1,2-dichloroethane, the two-step oxidation of **1** were observed at 0.52 and 0.80 V *vs.* SCE in 1,2-dichloromethane.
   Therefore, **1** and F_4TCNQ formed a completely ionized CT complex ($\mathbf{1}^+$)(F_4TCNQ^-).
9. M. B. Nielsen, C. Lomholt and J. Becher, Chem. Soc. Rev. **29,** 153 (2000); K. B. Simonsen, N Svenstrup, J. Lau, O. Simonsen, P. Mork, C. J. Kristensen and J. Becher, Synthesis, (1996) 407.
10. R. C. Wheland and E. L. Martin, J. Org. Chem **40,** 3101 (1975).
11. J. P. Torrance, J. J. Mayerle and K. Bechgaard, Phys. Rev. B. **22**, 4960 (1980).
12. C. S. Jacobsen, Optical Properties in Semiconductor and Semimetals. High Conducting Quasi-One-Dimensional Organic Crystals, edited by E. Conwell, (Academic Press, NY, 1988).
13. T. Akutagawa, Y. Abe, Y. Nezu, T. Nakamura, M. Kataoka, A. Yamanaka, K. Inoue, T. Inabe, C. A. Christensen and J. Becher, Inorg. Chem. **37**, 2330 (1998).
14. T. Akutagawa, Y. Abe, T. Hasegawa, T. Nakamura, T. Inabe, C. A. Christensen and J. Becher, Chem. Lett., 132 (2000).
15. T. Akutagawa, K. Kakiuchi, T. Hasegawa, T. Nakamura, C. A. Christensen and J, Becher, Langmuir **20,** 4187 (2004).
16. Y. Tatewaki, Y. Noda, T. Akutagawa, R. Tunashima, S. Noro, T. Nakamura, H. Hasegawa, S. Mashiko and J. Becher, Langmuir-Blodgett Films Constructed from a Charge-Transfer Complex and Gold Nanoparticles, J.Phys. Chem. C **111**, 18871 (2007).

Mater. Res. Soc. Symp. Proc. Vol. 1144 © 2009 Materials Research Society 1144-LL04-24

Structural and Electronic Properties of Rare-Earth Nanowires

Andrew Pratt, Charles Woffinden, Christopher Bonet, Steve P. Tear
Department of Physics, University of York, Heslington, York, YO10 5DD, U.K.

ABSTRACT

Sub-monolayer coverages of the rare-earth metal Ho were deposited onto Si(001) substrates at elevated temperatures resulting in the formation of rare-earth silicide nanowires. Between the nanowires the substrate area reconstructs into either a 2x4 or 2x7 reconstruction depending on the specific preparation conditions. We have studied the structural and electronic properties of both the nanowires and the reconstructed areas with the complementary techniques of scanning tunneling microscopy (STM) and metastable de-excitation spectroscopy (MDS) revealing the electronic similarities between the 2x4 and 2x7 phases. Evidence for the presence of hybridized Si $3s3p$-Ho $6s5d$ bonds suggests that these reconstructions form as a precursor to nanowire growth.

INTRODUCTION

Rare-earth (RE) metals deposited onto Si(001) shows promise as a materials system in which nanowires (NWs) naturally self-assemble [1]. This occurs due to an anisotropy in the lattice mismatch between the RE silicide formed after deposition at elevated temperatures and the underlying Si(001) substrate [2], and generally leads to NWs with widths and heights of several nm and lengths of up to 1 μm. Due to their potential application in nanoarchitectronics these structures have been extensively studied with NWs of the RE metals Sm, Gd, Dy, Ho and Er demonstrated, along with those of the chemically similar elements Sc and Y (see Ref. [3] and references therein). Despite this activity the structure of RE NWs is still to be completely solved. Much less attention has been paid to the study of the surface between the NWs although it is important to understand this region as it influences their initial formation and may act as a source of atoms for continued growth.

RE NWs formed on Si(001) are known to have a defected hexagonal-AlB_2 crystal structure and grow along the surface in a direction perpendicular to the dimer row orientation of the Si(001) 2x1 substrate (see Fig. 1). NWs form after the diffusion controlled reaction of the deposited RE with substrate Si atoms with the optimum substrate temperature during growth generally accepted to be around 600 °C. Depositing coverages greater than 1 ML leads to the formation of islands at the expense of NWs, as does higher deposition temperatures and longer anneal times. Between the NWs the surface has been observed to reconstruct into either a 2x4 or a 2x7 reconstruction, depending on the preparation conditions [4,5].

In this study, we have combined a variety of techniques to study the growth of RE NWs on Si(001) including scanning tunnelling microscopy (STM) and metastable de-excitation spectroscopy (MDS). MDS is of particular interest due to its extreme surface sensitivity with metastable atoms de-exciting several angstroms from the sample to yield an electron emission spectrum that reflects the surface density of states [6]. This surface sensitivity, along with the fact that the de-excitation mechanism is dependent on the conduction properties of the sample, provides information on the surface electronic and chemical properties with no contribution from the bulk, a problem in ultraviolet photoemission spectroscopy (UPS).

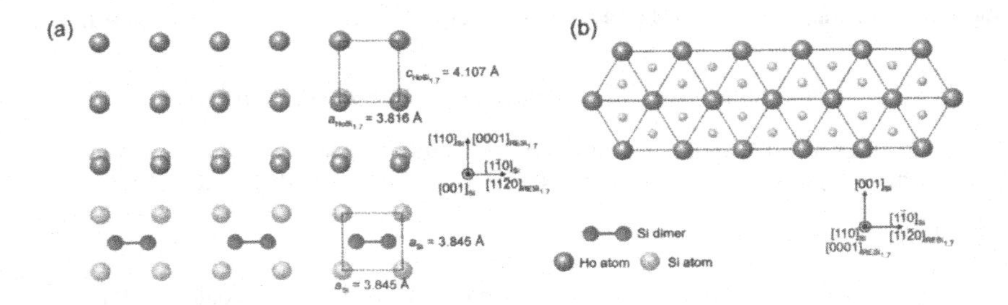

Figure 1. Schematic indicating (a) a plan view of the $HoSi_{1.7}$ geometry with respect to the Si(001) substrate, and (b) a side view of the hexagonal AlB_2-type structure of RE nanowires.

EXPERIMENTAL DETAILS

Samples were prepared by depositing sub-monolayer (ML) coverages of Ho onto Si(001) substrates that had been flash cleaned several times at 1200 °C followed by a slow cool to room temperature [3]. Low-energy electron diffraction (LEED) was used to confirm that a clean surface had been obtained before Ho deposition started. The substrate temperature was kept stable during deposition at between 500 and 650 °C and coverages between 0.1 and 1.0 ML were studied. STM and MDS experiments were carried out in separate vacuum systems operating at base pressures of <5 x 10^{-10} mbar. STM images were obtained using an Omicron Nanotechnology GmbH microscope. The MDS experiments were performed in a large system dedicated to the technique and described in detail in Ref. [7]. Electron energy spectra were obtained using an Omicron Vakuumphysik GmbH EA 125 hemispherical electron analyzer and to ensure good quality growth, a clean sample was used for each experiment.

DISCUSSION

Figure 2 (a) shows an STM image of Ho silicide NWs formed after 0.3 ML of Ho was deposited onto a clean Si(001) 2x1 surface with the substrate held at 530 °C during deposition and with no post-anneal. NWs on adjacent terraces grow orthogonally to each other as the growth occurs in a direction perpendicular to the Si substrate dimer row direction which rotates by 90° from step to step. NWs are terminated when they intersect with another NW or a barrier formed from bunched Si step edges. Depending on the preparation conditions (primarily the substrate temperature and Ho coverage), the surface between the NWs may reconstruct into either a 2x4 or 2x7 phase, or a combination of the two, and these are shown respectively in Fig. 2 (b) and Fig. 2 (c). The 2x4 phase generally appears before the onset of the 2x7 structure which is proposed to have a slightly higher Ho content [4,5]. The density of NWs, and the coverage at which the 2x7 reconstruction begins to dominate over the 2x4, was found to depend on the substrate temperature during deposition, as well as (more predictably) on coverage. A higher substrate temperature promotes Ho diffusion during deposition and the creation of NWs. This in turn leaves less Ho for surface reconstruction so that under these conditions the 2x4 phase will appear for higher coverages than for lower deposition temperatures.

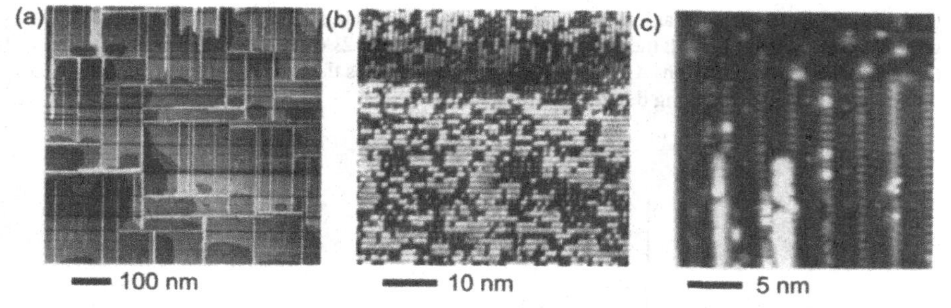

Figure 2. STM images showing (a) NWs formed after the deposition of 0.3 ML of Ho on Si(001) at a substrate temperature of 530 °C (700 x 700 nm^2, 2.0 V and 2.0 nA), (b) the 2x4 reconstruction (0.3 ML, 610 °C, 50 x 50 nm^2, -2.0 V, 2.0 nA) and (c) the 2x7 reconstruction (0.4 ML, 575 °C, 25 x 25 nm^2, -2.0 V, 2.0 nA).

The appearance of both the 2x4 and 2x7 reconstructions in the STM images (Figs. 2 (b) and 2 (c)) is characterized by the presence of distinctive maxima that disrupt the Si dimers present on the clean 2x1 surface (many dimers are still evident in Fig. 2 (b) for the 2x4 phase). A thorough STM study has been carried out on both the 2x4 and 2x7 surfaces leading to a model proposing that the maxima are due to the presence of RE atoms at the outermost layer of these reconstructions [4,5]. However this deduction is based on the registry of atomic features in the acquired images with the substrate lattice, and on calculations of metal-atom conservation. With STM alone, the idea that the surface of these reconstructions is composed of Si atoms, with the deposited RE atoms sinking below, cannot be ruled out. In fact, evidence for this suggestion comes from the well-studied case of RE deposition on Si(111) in which Ho layers form below a buckled Si surface bilayer [8].

To try and identify the features of the 2x4 and 2x7 reconstructions, as observed by ourselves and others with STM, MDS was conducted on these surfaces [3]. Figure 3 shows the spectra resulting from a variety of coverages and from samples which were prepared by depositing at a substrate temperature of 500 °C—slightly lower than is optimum for NW growth in order to promote surface reconstruction. The electronic properties of samples with different coverages of Ho deposited onto them, as measured here using MDS, can be assigned to specific reconstructions by comparison with LEED data which was used to identify the surface structure, as indicated in the figure.

On the Si(001) 2x1 surface He 2^3S de-excitation is known to occur through a two-stage process—resonance ionization (RI) followed by Auger neutralization (AN) [4]. RI occurs when the excited 2*s* electron of the metastable He atom tunnels into an empty degenerate state of the surface leaving a He^+ ion. This is then neutralized by an electron from the surface valence band which tunnels into the ground state of the He atom with the excess energy liberating a second electron from the surface in an Auger-type process (AN). As the energy of the emitted electron is determined by two electrons, the spectra resulting from surfaces that de-excite via this mechanism are generally broad and featureless, as seen in Fig. 3. If the surface under study is insulating, or if it has a very low work function then an alternative process to RI+AN known as Auger de-excitation (AD; equivalent to Penning ionization) may occur [4]. In this instance there is no state in the surface degenerate in energy with the 2*s* electron of the metastable He atom so that RI is inhibited. Instead, an electron from the surface tunnels directly into the half-filled ground state of the atom with the simultaneous ejection of the 2*s* electron in a process reminiscent of photoemission. As AD is a one-electron process the resulting spectra are easier to interpret with bands appearing due to specific features.

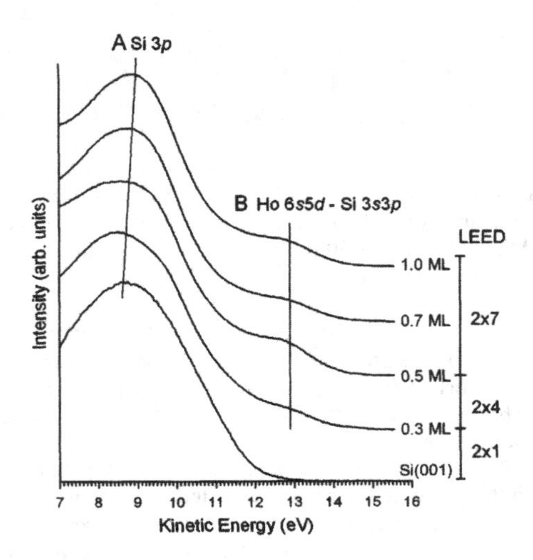

Figure 3. MDS spectra for sub-ML coverages of Ho deposited onto a clean Si(001) 2x1 substrate held at 500 °C with the clean Si(001) 2x1 spectrum shown for comparison.

In Fig. 3, the increased emission in the spectra obtained from Ho-reconstructed surfaces at kinetic energies greater than the cut-off for the clean Si(001) surface (in the range 12-14 eV) suggests that the de-excitation channel has changed from RI+AN to AD. This indicates that the 2x4 and 2x7 reconstructions are low work-function surfaces which may be explained by the presence of Ho-Si bonds. Charge is transferred from Ho atoms to Si atoms resulting in a lowering of the work function through the surface dipole effect [4]. The lack of emission at the Fermi level cut-off at around 14.5 eV lends evidence to the suggestion that the Ho/Si(001) interface is semi-metallic [9].

Two main features appear in Fig. 3, labeled A and B. The former is attributed to emission from Si 3*p* states which are present both at the surface and in the bulk of clean Si(001) 2x1, but also in the 2x4 and 2x7 reconstructed phases. The environment surrounding the Si atoms is different for the two cases though resulting in a slight shift in this peak as the coverage increases and the area of clean surface reduces. This is supported by a photoemission study of the Er/Si(001) interface which showed a core-level shift of approximately 0.3 eV for the Si 3*p* state after the deposition of 0.6 ML of Er [10]. Feature B is assigned to emission from bonds between hybridized Si 3*s*3*p* and Ho 6*s*5*d* states present at the surface of both the 2x4 and 2x7 reconstructions. This assertion is based on a comparison with an MDS study of high coverages of Er and Yb on Si(001) [11] and by comparison with partial DOS calculations for Er silicide [12].

CONCLUSIONS

MDS has revealed the presence of hybridized Si 3*s*3*p* – Ho 6*s*5*d* bonds at the surface of the 2x4 and 2x7 reconstructions that form between NWs when sub-ML coverages of Ho are deposited onto clean Si(001) substrates. This suggests that prominent maxima observed in corresponding STM images are due to the presence of Ho atoms at the surface of these reconstructions—an uncertainty before now—and that these form as a precursor to NW growth. This information will help in the deduction of a starting model for structural derivation techniques such as LEED *I*(*V*) and/or *ab initio* calculations.

ACKNOWLEDGMENTS

We would like to acknowledge the Engineering and Physical Sciences Research Council (U.K.) for funding this research.

REFERENCES

1. D. Bowler, *J. Phys.: Condens. Matter* **16**, R721 (2004).
2. C. Ohbuchi, J. Nogami, *Phys. Rev. B* **66**, 165323 (2002).
3. A. Pratt, C. Woffinden, C. Bonet and S.P. Tear, *Phys. Rev.* B **78**, 155430 (2008).
4. B.Z. Liu and J. Nogami, *Surf. Sci.* **540**, 136 (2003).
5. B.Z. Liu and J. Nogami, *Surf. Sci.* **488**, 399 (2001).
6. Y. Harada, S. Masuda and H. Ozaki, *Chem. Phys.* **97**,1897 (1997).
7. A. Pratt, A. Roskoss, H. Ménard and M. Jacka, *Rev. Sci. Instr.* **76**, 053102 (2005).
8. C. Bonet, I. M. Scott, D. J. Spence, T. J. Wood, T. C. Q. Noakes, P. Bailey and S. P. Tear, *Phys. Rev.* B **72**, 165407 (2005).
9. R. Hofmann, W. A. Henle, F. P. Netzer and M. Neuber, *Phys. Rev.* B **46**, 3857 (1992).
10. G. Chen, X. Ding, Z. Li and X. Wang, *J. Phys.: Condens. Matter* **14**, 10075 (2002).
11. L. Pasquali and S. Nannarone, *Nucl. Instrum. Methods Phys. Res.* B **230**, 340 (2005).
12. L. Magaud, J. Y. Veuillen, D. Lollman, T. A. Nguyen Tan, D. A. Papaconstantopoulos and M. J. Mehl, *Phys. Rev.* B **46**, 1299 (1992).

Mater. Res. Soc. Symp. Proc. Vol. 1144 © 2009 Materials Research Society 1144-LL06-02

Effect of Metal-Silicon Nanowire Contacts on the Performance of Accumulation Metal Oxide Semiconductor Field Effect Transistor

Pranav Garg[1], Yi Hong[1], Md Mash-Hud Iqbal[2], and Stephen J. Fonash[1]

1. Center for Nanotechnology Education and Utilization, The Pennsylvania State University, University Park, PA, USA.
2. Centre for Advanced Photonics and Electronics, University of Cambridge, Cambridge, United Kingdom.

ABSTRACT

Recently, we have experimentally demonstrated a very simply structured unipolar accumulation-type metal oxide semiconductor field effect transistor (AMOSFET) using grow-in-place silicon nanowires. The AMOSFET consists of a single doping type nanowire, metal source and drain contacts which are separated by a partially gated region. Despite its simple configuration, it is capable of high performance thereby offering the potential of a low manufacturing-cost transistor. Since the quality of the metal/semiconductor ohmic source and drain contacts impacts AMOSFET performance, we repot here on initial exploration of contact variations and of the impact of thermal process history. With process optimization, current on/off ratios of 10^6 and subthreshold swings of 70 mV/dec have been achieved with these simple devices.

INTRODUCTION

Field effect transistors (FET) based on semiconductor nanowires (NW) have received enormous attention due to their potential applications as building blocks for future nanoscale electronics [1-4]. Many of these NW FET devices are fabricated on single doping type NW with metallic source/drain (S/D) Schottky barriers (SB) contacts and a back gate. Many function as accumulation channel devices but show strong ambipolar conduction [2, 3], large hysteresis [4], and current-voltage (I-V) characteristics that are limited by S/D Schottky barrier contacts [3, 4]. Efforts to suppress ambipolar behavior and improve I-V characteristics such as double gate structures [3] and/or S/D contact doping [4] require additional processing steps and face the challenges of channel length scaling. The AMOSFET offers a very simple device configuration that can give high performance unipolar transistors (on nanowires, nanoribbons or thin films) while avoiding complex processing steps.

Figures 1a and 1b schematically show the AMOSFET configurations on a semiconductor thin film and on a nanowire, respectively. As shown, the AMOSFET configuration only requires a single doping type semiconductor as the active layer, ohmic source and drain contacts spaced at minimum required distance from the gate, a minimum length gate, and a nano-scale dimension perpendicular to the gate [5-8]. Extensive numerical simulations of the AMOSFET configuration have shown that the nanoscale dimension perpendicular to the gate required in AMOSFETs makes it possible for the gate to control on-state current conduction via accumulated majority carriers and off-state behavior through depletion of majority carriers [5,6]. The gate dielectric thickness and dielectric constant are not critical since the AMOSFET I-V characteristics have a very weak dependence on gate capacitance, and the use of a minimum gate length suppresses the possibility of ambipolar conduction [5,6]. In this report we use AMOSFETs on silicon nanowires (SiNW) with a structure shown in Figure 1b [7,8]. The device uses a p-doped SiNW as active layer, a thermally grown oxide wrapping around the SiNW employed as gate dielectric, a metal gate surrounding the dielectric, and metal S/D contacts. We now examine the impact of various metals for these S/D contacts and of thermal processing history on characteristics of this AMOSFET on Si nanowires..

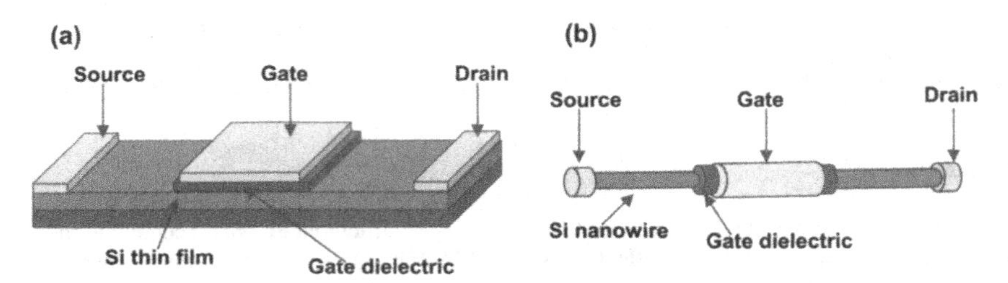

Figure 1: Schematic view of AMOSFET configuration on (a) a semiconductor thin film, and on (b) a semiconductor nanowire.

EXPERIMENTAL PROCEDURE

The AMOSFET devices fabricated for this work used nominally undoped SiNWs with diameters of ~100 nm synthesized by the gold catalyzed vapor-liquid-solid (VLS) mechanism in a low pressure chemical vapor deposition (LPCVD) system using our "grow-in-place" approach [7,8]. "Grow-in-place" uses predefined nanochannel templates with a VLS catalyst to allow easy control of the number, position and orientation of the nanowires, relieving the need of collection, positioning and assembling steps in making nanowire devices. The resulting SiNWs may be extruded from the template or confined to the template and, in either case, subsequently fabricated into devices. In the case of this report, we used extruded SINWs to fabricate the transistors.

The resulting as-grown SiNWs, which are p-type [7,8], were cleaned before dry oxidation at 700 °C for 4 hours which results in a gate oxide of about 10 nm. FETs and four point probe structures were fabricated simultaneously, to characterize device performance and bulk/contact properties, respectively. Four point probe structure contacts and FET S/D contacts were defined by photolithography followed by oxide etching, metal deposition and lift-off. Two different work function metals, low work function titanium (Ti) and high work function cobalt (Co) were used for the probe and S/D contacts. For FETs, an additional lithography step was needed to define a Ti gate electrode between the S/D contacts. Figures 2a and 2b show a schematic view and field emission scanning electron microscope (FESEM) image, respectively, of a four point probe structure fabricated on a grow-in-place SiNW. Figures 2c and 2d show a schematic view and FESEM image, respectively, of an AMOSFET device fabricated on a grow-in-place SiNW.

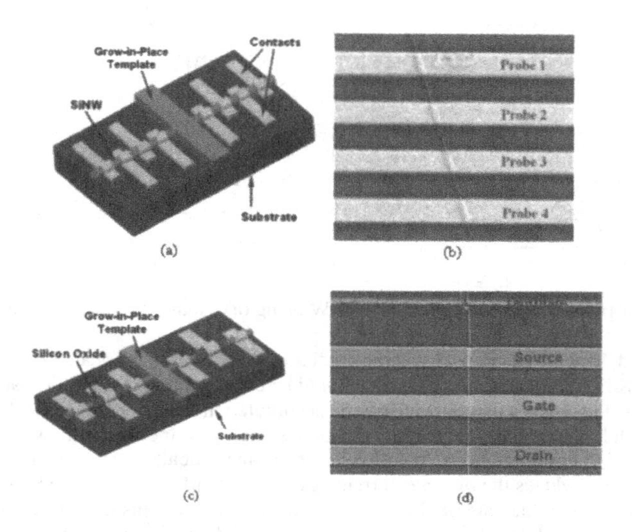

Figure 2: (a) Schematic view and, (b) top FESEM view of four point probes fabricated on a grow-in-place SiNW. Also shown are the (c) schematic view and, (d) Top FESEM view of AMOSFETs fabricated on a grow-in-place SiNW. The red bar in micrographs corresponds to 2 µm.

RESULTS AND DISCUSSION

Figures 3a and 3b show the results of four point I-V measurements using titanium and cobalt contacts, respectively. In these measurements, the voltage between probes 1 and 4 (V_{14}) is swept while, the voltage drop between probes 2 and 3 (V_{23}) as well as the current through from probe 1 to 4 (I_{14}) are measured. The slope for V_{23} vs. I_{14} (pink curve) gives the "bulk" resistance of the SiNW segment between probes 2 and 3 (R_{23}) independent of contacts effects. The contact resistance R_c can then also be estimated. Care must be taken in calculating the contact resistance since current can shunt across the metallization at probes 2 and 3 due to the low conductance of undoped Si nanowire. For titanium contacts (from Figure 3a), R_{23} and R_c are estimated to be 30×10^7 ohms, and 1.45×10^9 ohms, respectively. For cobalt contacts (from Figure 3b), R_{23} and R_c are estimated to be 11×10^7 ohms, and 1.5×10^7 ohms, respectively. Note that the SiNW resistance is of the same order of magnitude in both cases. However, the contact resistance for titanium contacts is two orders of magnitude higher than that for cobalt which correlates with the barrier heights expected for these metals on "bulk" p-type silicon [9]. Using values of R_{23} and the dimensions of the SiNW, the bulk resistivity (ρ_{SiNW}) of the nanowires can also be estimated as 120 Ω-cm and 40 Ω-cm, in case of titanium and cobalt contacts, respectively. These values agree well with previously published results on unintentionally doped VLS grown SiNW [8, 10]. For bulk p-type silicon, these resistivity values correspond to doping levels of ~ $10^{14} – 10^{15}$ cm^{-3} [9]. However, due to differences in the mobility of holes in nanowire compared to bulk silicon, the actual doping level in the SiNW may be very different from bulk values.

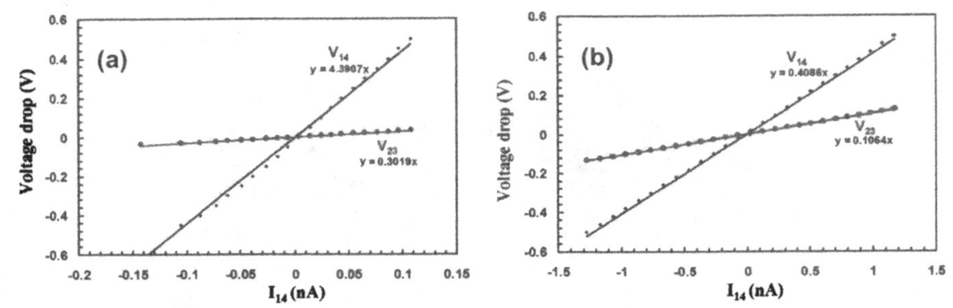

Figure 3: Four point I-V characteristics of SiNW using (a) Titanium, and (b) Cobalt contacts.

Figures 4a and 4b show the transfer characteristics of AMOSFETs fabricated using titanium and cobalt as the S/D contacts, respectively. As would be expected for AMOSFETs fabricated using p-type SiNWs, the devices demonstrate p-type accumulation behavior. Also, the devices fabricated with higher resistance titanium contacts exhibit extremely poor transfer characteristics when compared to devices fabricated with lower resistance cobalt contacts. Higher S/D series contact resistance reduces the on-state current of the device. Moreover, as the S/D series contact resistance is increased, the gate field control over the channel diminishes. Hence the AMOSFET behavior moves toward that of a resistor, thereby causing the degradation of subthreshold characteristics.

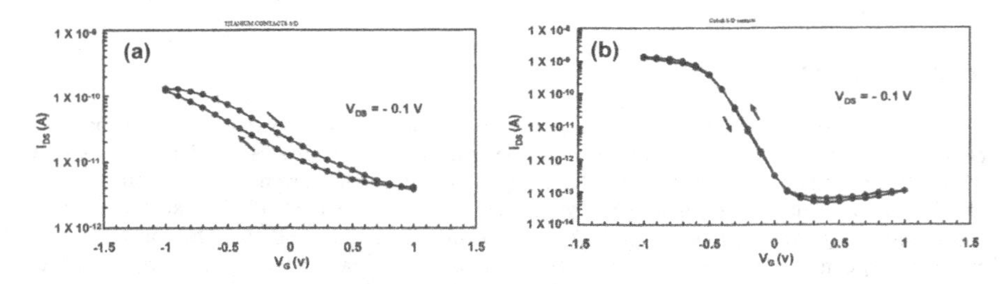

Figure 4: Transfer characteristics of AMOSFETs fabricated using (a) Titanium, and (b) Cobalt S/D contacts. Titanium is used as the gate metal in both cases. The On/Off current ratio and subthreshold swing (*SS*) for titanium S/D contacted devices are ~ 30 and ~ 1000 mV/dec, respectively. The On/Off current ratio and subthreshold swing for these cobalt S/D contacted devices are ~ 10^4 and ~ 150 mV/dec, respectively.

The performance of the AMOSFETs with cobalt S/D contacts seen in Fig. 4 can be significantly improved by annealing the devices at temperatures between 300 – 500 °C in inert nitrogen or forming gas atmospheres. Annealing helps to improve S/D contacts by improving adhesion as well as promoting formation of low barrier silicide phases [1, 3].Figure 5a shows an example of such improvements in transfer characteristics after furnace annealing in nitrogen atmosphere at 350 °C for 10 - 15 minutes. After annealing, the subthreshold swing reduces from 150 mV/dec to 100 mV/dec and the on/off current ratio increases from 10^4 to 10^6. Furnace annealing at higher temperature and/or longer periods resulted in failed devices. However, it was possible to anneal the devices at higher temperature but for very short period using rapid thermal annealing (RTA). Figure 5b shows an example of improvements in transfer characteristics after RTA in forming gas (10% hydrogen in nitrogen) atmosphere at 500 °C for 30s. After annealing, the subthreshold swing reduces from 250 mV/dec to 70 mV/dec and the on/off current ratio increases from 10^3 to 10^6. These values are some of the best values reported in literature for SiNW FET devices.

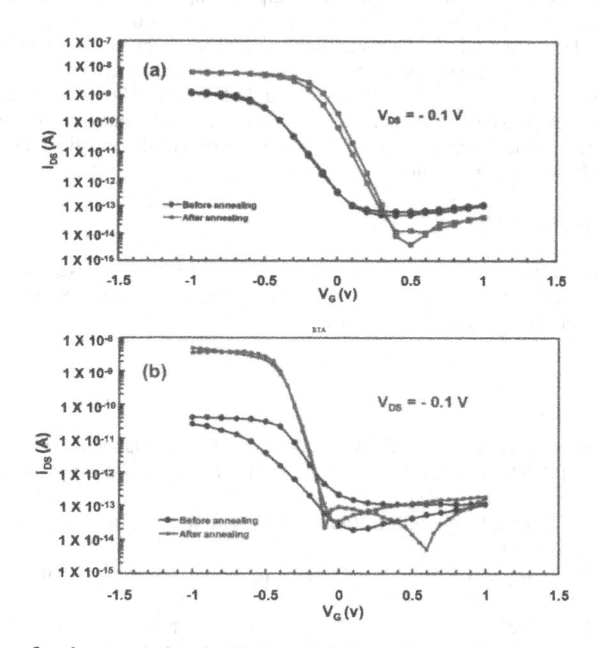

Figure 5: (a) Transfer characteristic of a SiNW AMOSFET using cobalt S/D contacts before and after furnace annealing at 350 °C for 15 mins in N_2 atmosphere. (b) Transfer characteristic of a SiNW AMOSFET using cobalt S/D contacts before and after RTA at 500 °C for 30 s in forming gas (10% H_2 in N_2) atmosphere.

SUMMARY

We have further studied the high-performance AMOSFET transistor configuration which we have reported on earlier [5-8]. The configuration is suitable for fabrication on thin film, nanowire, and nanoribbon semiconductors. Here, AMOSFET devices were fabricated on SiNWs grown using our "grow-in-place" approach [7,8]. Two different metals, low work function titanium and high work function cobalt, were investigated to determine the impact of their use as S/D contacts. It has been found that titanium forms extremely high resistance contacts with our SiNWs which results in very poor AMOSFET characteristics. Devices fabricated using cobalt S/D contacts exhibited much superior performance. Thermal annealing of devices with Co S/D contacts using furnace or rapid thermal annealing helps to improve these contacts and device performance. After this annealing, device behavior with on-off ratios of 10^6 and sub-threshold swings of 70 mV/dec were attained. This examination demonstrates that the AMOSFET structure can be robust to thermal process and further elaborates that high performance devices can be attained using the extremely simply structured AMOSFET. The experimental results presented clearly support what our earlier experimental and simulation work has predicted; i.e., AMOSFET devices constructed using thin film, nanowire or naoribbon silicon hold the promise of low cost manufacturing yet excellent performance.

ACKNOWLEDGEMENTS

The work was supported in part by NSF Grant No. DMI-0615579. This research made extensive use of the facilities of the Penn State Nanofabrication Facility, a node of the NSF National Nanotechnology Infrastructure Network

REFERENCES

1) Y. Cui, Z.H Zhong, D.L. Wang, W.U. Wang and C.M. Lieber, Nano Letters **3**, 149 (2003).
2) S.M. Koo, M.D. Edelstein, Q. Li, C.A. Richter and E.M. Vogel, Nanotechnology **16**, 1482 (2005).
3) J. Appenzeller, J. Knoch, E. Tutuc, M. Reuter and S. Guha, 2006 International Electron Devices Meeting, 1 (2006).
4) Y. Wang, K.-K. Lew, J. Mattzela, J. M. Redwing and T. S. Mayer, 63rd Device Research Conference Proceedings **1**, 159 (2005).
5) S. J. Fonash, M. M. Iqbal, F. Udrea, and P. Migliorato, Applied Physics Letters **91**, 193508 (2007).
6) M. M. Iqbal, Y. Hong, P. Garg, F. Udrea, P. Migliorato, and S. J. Fonash, IEEE Transactions on Electron Devices **55**, 2946 (2008).
7) Y. Shan, S. Ashok, and S. J. Fonash, Applied P

hysics Letters **91**, 093518 (2007).
8) Y. Shan and S. J. Fonash, ACS Nano **2**, 429 (2008).
9) S. M. Sze and Kwok K. Ng, *Physics of* Semiconductor *Devices*, 3rd ed. (John Wiley & Sons, Inc., Hoboken, New Jersey, 2007) p. 179.
10) Y. Wang, M. Cabassi, T. Ho, K-K Lew, J. Redwing, and T. Mayer, 62nd Device Research Conference Proceedings **1**, 23 (2004).

Mater. Res. Soc. Symp. Proc. Vol. 1144 © 2009 Materials Research Society 1144-LL06-04

High Performance Printed Aligned Carbon Nanotube Transistors on Both Rigid and Flexible Substrates for Transparent Electronics

Hsiao-Kang Chang, Fumiaki N. Ishikawa, Koungmin Ryu, Pochiang Chen, Alexander Badmaev, Guozhen Shen, and Chongwu Zhou

Department of Electrical Engineering, University of Southern California, Los Angeles, CA 90089, USA

ABSTRACT

We report high-performance fully transparent thin-film transistors (TTFTs) on both rigid and flexible substrates. Such transistors have been fabricated through low temperature processing, which allowed device fabrication even on flexible substrates. Transparent transistors with high effective mobilities (~1,300 $cm^2V^{-1}s^{-1}$) were first fabricated through engineering of the source and drain contacts. High on/off ratio (3×10^4) was achieved via electrical breakdown. Flexible transparent TTFTs were fabricated and successfully operated under bending up to 120°. The transparent transistors were further utilized to construct a fully transparent and flexible logic inverter, and also used to control commercial GaN light–emitting diodes (LEDs) with a light intensity modulation of 10^3. The aforementioned results suggest that aligned nanotubes have great potential as building blocks toward future transparent electronics.

INTRODUCTION

Transparent electronics is an emerging technology that has attracted numerous research efforts in recent years due to its potential to make significant impact in a wide variety of areas.[1-4] The development of transparent thin-film transistors (TTFTs) is crucial for transparent electronics. The key performance criteria of TTFTs are high device mobility and low temperature fabrication. High device mobility enables fast device operation and low power consumption, while low temperature fabrication is essential to transparent devices made on flexible substrates.

Wide band-gap semiconductor films[1, 2] and nanowires[4] have been studied for TTFTs. However, the reported mobility values are still lower than non-transparent devices. In addition, oxide based TTFTs were limited to n-type transistors.[1, 2, 4] The development of high performance transparent p-type transistors, which is an essential element in CMOS technology, remains a great challenge.

To achieve high performance p-type TTFTs, single-walled carbon nanotube (SWNT) is a promising candidate with intrinsic mobility over 100,000 $cm^2V^{-1}s^{-1}$,[5] good mechanical flexibility,[6] p-type transport behaviour and good optical transparency.[3] Recently, random nanotube networks were used as active channels for TTFTs[3] with highest obtained mobility around 30 $cm^2V^{-1}s^{-1}$.[3] This low mobility might have resulted from the fact that electrical conduction in a random nanotube network has to go through many nanotube-nanotube junctions.

Aligned carbon nanotubes which can directly bridge source and drain are therefore expected to offer better performance.

In this paper, we introduce both high mobility and low-temperature processing for TTFTs made with highly aligned single-walled carbon nanotubes. As opposed to random networked nanotubes, the use of massively aligned nanotubes enables the devices to exhibit good performance such as high mobility, good transparency and mechanical flexibility. These aligned nanotube transistors are easy to fabricate and integrate, as compared to individual nanotube devices. The transfer printing technique allows the devices to be fabricated through low temperature process, which is particularly important for realizing transparent electronics on flexible substrates.

EXPERIMENTAL DETAILS

Figure 1a shows a schematic diagram of our transfer printing method and the structure of the transparent transistors. The aligned nanotubes were first grown on quartz substrates using chemical vapour deposition (CVD). The transfer started by coating as-grown aligned SWNTs on a quartz substrate with a 100 nm thick Au film. Revalpha thermal tape (from Nitto Denko), which is adhesive at room temperature but loses adhesion above 120 C^{o}, was placed on the Au film, and then peeled off slowly that resulted in picking up the nanotube/Au film. The thermal tape/Au film/aligned SWNT film was then placed onto a target substrate with an ITO back gate and a SU8 dielectric layer, and the whole structure was heated up to 130 C^{o} on a hot plate to detach the thermal tape. The Au film was removed by a gold etchant, leaving the aligned SWNTs on target substrate. Finally, source and drain electrodes made of thin layer of Au plus ITO or ITO only were defined by photolithography and lift-off techniques.

The printing transfer method allowed the construction of devices on any substrates including flexible ones, and a device yield as high as 100% was achieved due to uniform coverage of the substrates with the aligned SWNTs. Shown in Figure 1b is a representative scanning electron microscopy (SEM) image of transferred aligned SWNTs. Most of nanotubes can directly bridge the source and drain for the device dimensions studied. Figure 1c is a SEM image of the devices with different channel widths from 8 μm to 100 μm. The channel length keeps at 4 μm. Inset in Figure 1c is a typical SEM image of the aligned SWNTs between source and drain, showing nanotubes remain highly ordered after the whole fabrication process. Figure 1d and 1e depict the optical micrographs of devices fabricated on a 4-inch glass wafer and a piece of PET sheet, respectively. They show good transparency as the backgrounds can be easily seen through these devices.

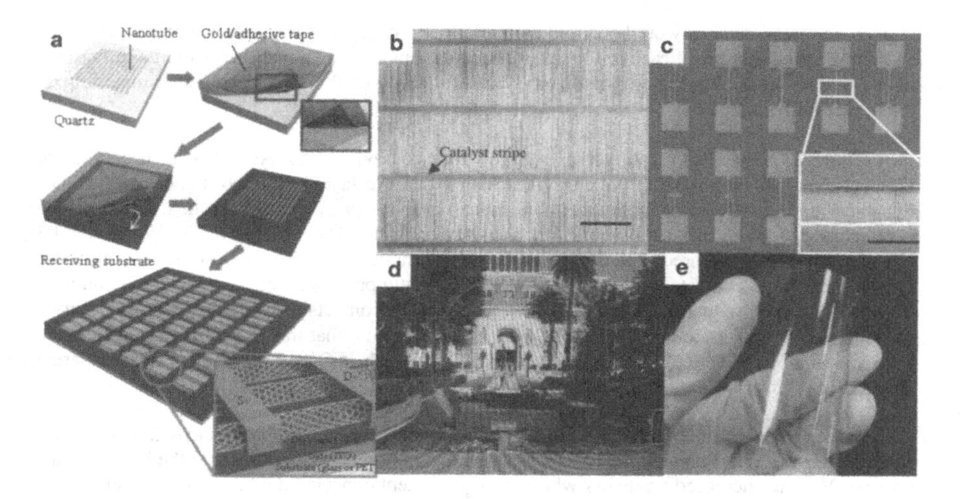

Figure 1. a, Schematic diagram of aligned SWNT transfer and a device structure. b, SEM image of transferred aligned SWNTs (Scale bar: 50 μm) c, SEM image of aligned nanotube transistors Inset: SEM image of aligned nanotubes bridging ITO electrodes. (Scale bar: 5 μm). d, e. Optical micrographs of arrays of aligned nanotube transparent transistors on a 4-inch glass wafer and on a 3" by 4" PET sheet, respectively.

DISCUSSION

To fabricate transparent transistors, we used ITO film on glass as the back gate, 2 μm thick SU8 as gate dielectric and two kinds of source/drain contact materials: 1 nm Au/100 nm ITO and 100 nm ITO for device fabrication. These transistors have aligned nanotubes as active channel with channel width of 10, 20, 50, 100, and 200 μm and channel length of 4, 10, 20, 50, and 100 μm. The transparency of the devices, shown in Figure 2a, indicates that the devices with Au/ITO contacts had a transmittance of ~80% in visible light regime. Compared to the devices using only ITO as contacts, the transmittance decreased about 5 % due to thin Au layer.

The performance of our devices was then characterized. Figure 2b and 2c are representative plots of drain-source current (I_{ds}) versus drain-source voltage (V_{ds}) and I_{ds} versus gate voltage (V_g) for a device with a channel length of 50 μm and a width of 100 μm, respectively. The device showed p-type transistor behavior and an on/off ratio ~20 due to the presence of metallic nanotubes. The large operation voltage can be easily improved by utilizing high-k dielectrics.

Figure 2d shows mobility plotted against channel length. An effective mobility of 1,300 $cm^2V^{-1}s^{-1}$ was realized for the devices with channel length of 100 μm. This mobility is the highest one among transparent transistors using various active materials reported so far.[2-4] Channel length dependent mobility was clearly observed, indicating the presence of small Shottky barriers between the carbon nanotubes and the Au/ITO contacts.[7] The high mobility of aligned nanotube devices would enable low operation voltage, low power consumption, and high switching speed that are attractive for several applications such as transparent circuits in portable displays. This high mobility was realized by careful study of the effect of contact materials on the device performance.

Figure 2e shows that the I_{ds} versus V_{ds} plots for two devices with Au/ITO (linear curve) and ITO only contact (nonlinear curve). Devices with a thin Au layer exhibited higher conductance by about three orders of magnitude than those without Au. We attribute this difference in conductance to the difference in work functions for gold and ITO, which are 5.3 eV and 3.9~4.4 eV, respectively. Compared with the nanotube work function (4.7~5.1 eV[8]), we expect gold can form rather ohmic contacts to nanotubes, while ITO would present a Schottky barrier at contacts. This is consistent with the observation that the Au/ITO contacts led to a linear I_{ds}-V_{ds} curve, while the ITO contact yielded nonlinear I_{ds}-V_{ds} curve. We note that this is the first report of using a thin layer of gold to reduce the Schottky barrier between ITO and nanotubes for transparent electronics.

Improvement of the on/off ratio for the transparent aligned nanotubes devices can be further achieved via electrical breakdown to damage metallic carbon nanotubes. During the breakdown process, V_{ds} was increased gradually while the V_g was kept constant at a high positive value (20 V). Figure 2f shows a family of I_{ds}-V_{ds} curves of a device after breakdown. Gate voltages with steps of 5 V were applied. The results indicate good gate dependence and on/off ratio with clearly separated curves. Inset of Figure 2f is the I_{ds}-V_g curve in log scale with $V_{ds} = 0.5$ V, showing an on/off ratio of 3×10^4.

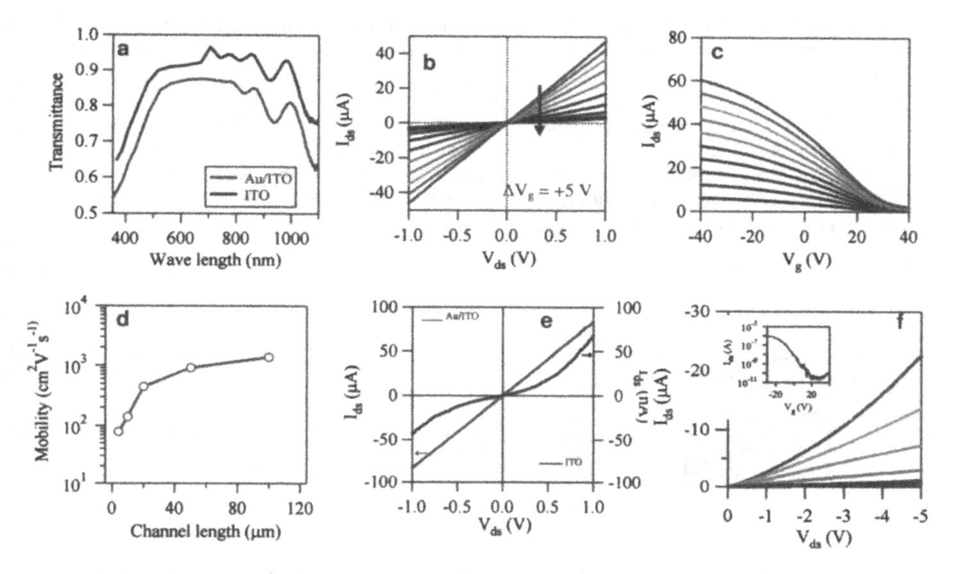

Figure 2. a, Optical transmittance of glass substrates with arrays of transparent aligned SWNT transistors with Au/ITO (lower curve) and ITO (higher curve) contacts. b, I_{ds}-V_{ds} curves of a 100 um wide and 50 um long aligned SWNT transistor with Au/ITO contact under different V_g from -20 V to 20 V. c, I_{ds}-V_g curves of the same device under different V_{ds} from 0.1 to 1.0 V. d, Mobility versus channel length for aligned SWNT transistors with Au/ITO contacts. The channel width of all the devices was 200 μm. e, Typical I_{ds}-V_{ds} plots for devices Au/ITO (linear curve) and ITO (nonlinear curve) contacts under V_g = 0 V, respectively. f, I_{ds}-V_{ds} curves of an aligned SWNT transistor showing high on/off ratio. V_g varied from -10 V to 10 V with step of 4 V. Inset shows an I_{ds}-V_g curve of the same device in log scale with V_{ds} =0.5

Performance of Aligned SWNT flexible TTFTs

Fully transparent and flexible aligned SWNTs devices using PET substrates were also fabricated with ITO back gate, SU8 gate dielectric and ITO source and drain (Figure 1a). Electronic devices on flexible substrates are extremely attractive owing to the emergence of wearable, handheld, portable consumer electronics as well as the compatibility with roll-to-roll fabrication.[9] Figure 3a is the transmittance of the transparent and flexible devices. The optical transmission is ~80 % in the 350-1200 nm wavelength range. Inset of Figure 3a is an optical micrograph of devices on a piece of PET sheet.

To evaluate the flexibility of our devices, I_{ds}-V_g measurements were performed under bending of the substrates with different angles and the plots are shown in Figure 3b. The I_{ds}-V_g curves correspond to a device bended for 0°, 30°, 60°, 90°, and 120° with a channel length of 10 μm and width of 200 μm. One can easily notice that the device continued to perform under bending angles from 0° to 120°, and the variation for both transconductance and on current were rather small. Inset of Figure 3b shows a picture of the bending experimental setup with bending angle defined as 180° - θ.

Furthermore, the device was successfully operated as a transistor on a bended substrate as shown in Figure 3c, where I_{ds}-V_{ds} curves at different V_g were plotted. The aforementioned data shows that aligned nanotube transparent transistors were successfully fabricated on PET substrates with good mechanical flexibility. One can improve the transconductance easily by depositing a thin Au layer between SWNTs and ITO, as demonstrated above.

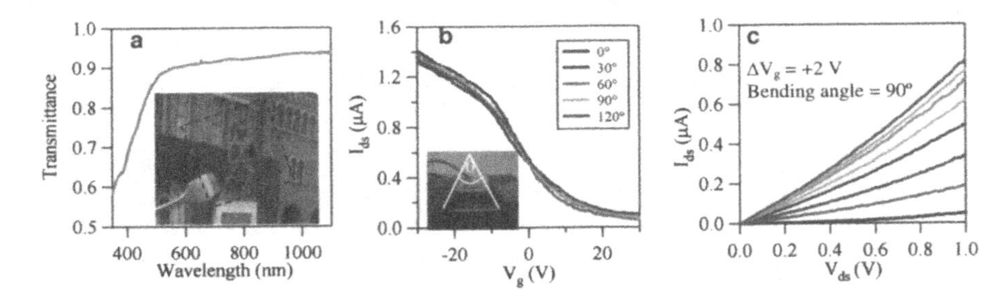

Figure 3. a, Optical transmittance of a PET substrate with arrays of transparent aligned SWNT transistors with ITO contacts. Inset is an optical micrograph of a piece of PET sheet with aligned nanotube TTFTs. b, I_{ds}-V_g curves of a representative device under different bending angles. Inset shows an optical micrograph of the experimental setup to measure I_{ds}-V_g under bending of the substrate. c, I_{ds}-V_{ds} curves of a fully transparent flexible aligned SWNT transistor under different V_g with bending of the substrate by 90° degree.

Transparent Flexible PMOS Inverters and LED Driving Circuitry

Our ability to fabricate high performance transparent and flexible transistors enabled us to apply them toward transparent circuits and transparent displays. Figure 4a shows an optical micrograph of fully transparent and flexible PMOS logic gates (inverters) on PET using transfer printed aligned SWNTs. The inset is the circuit diagram of the inverter, where one aligned SWNT transistor with an individually addressable gate was used as the drive, and another transistor with individual back gate fixed at V_g = 0 V as the load. The output voltage (V_{out}) of the inverter was plotted versus V_{in} in Figure 4b, together with the gain defined as dV_{out}/dV_{in}. A maximum gain of ~0.38 was obtained at V_{in} = ~ 12 V. Gain of the inverter is related to the polymer dielectric used, and can be improved by using high-k dielectric and/or thinner dielectric materials.

Finally, we have studied using aligned nanotubes transparent transistors to control various light emitting devices as a proof-of-concept for future transparent displays. A GaN LEDs was selected for the study and was wire-bonded on a breadboard with aligned nanotube TTFTs. Figure 4c shows the circuit diagram of this experiment, where one TTFT was connected to LED, and V_{DD} was applied to drain of the transistor. By controlling V_{in} that worked as gate voltage for the transistor with fixed V_{DD}, we can control the voltage drop across the LED. By applying different V_{in}, the current flowing through LED was successfully modulated by a factor of ~7. This modulation led to control of LED light intensity as shown in Figure 4d, where the light intensity was plotted against V_{in} in linear (left) and in log (right) scale, respectively. The ratio of the light intensity at on state and off state reached ~10^3. Figure 4e shows photographs of the LED operated with V_{in} = 30 (left), 0 (middle), -30 (right) V, respectively. The image clearly shows significant modulation in light intensity.

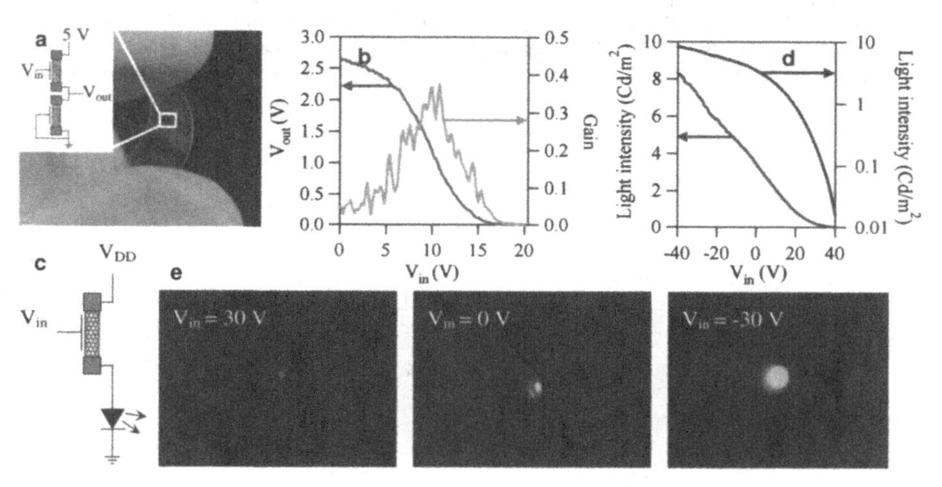

Figure 4. a, Optical micrograph of fully transparent and flexible logic gates (inverters). Inset shows the circuit diagram of the PMOS inverter. b, Plots of V_{out} and inverter gain versus V_{in} of a fully transparent flexible PMOS inverter. c, The circuit diagram of a LED driven by a transparent SWNT transistor. d, LED light intensity versus V_{in} in linear (left axis) and log (right axis) scale with V_{DD} = 9 V. f, Optical images of the LED under V_{in} = 30 V, 0 V, and -30 V.

CONCLUSIONS

In summary, we have fabricated aligned single-walled carbon nanotubes TTFTs for high-performance transparent electronics. These aligned nanotube transparent transistors showed good transmittance, very high mobility, and can be made compatible with flexible substrates due to the low-temperature processing employed. We achieved a device mobility of 1,300 $cm^2V^{-1}s^{-1}$, which is the highest among transparent transistors reported so far. Electrical breakdown was then utilized to obtain high on/off ratios (~3×10^4). Fully transparent and flexible transistors were also fabricated, showing very good mechanical flexibility as transistors worked properly under bending up to 120°. The transparent transistors were utilized to construct a fully transparent and flexible logic inverter on a plastic substrate, and also used to control commercial GaN light-emitting diodes with a light intensity modulation up to 10^3. Our work indicates that aligned nanotubes have great potential toward high-performance transparent and flexible electronics.

REFERENCES

1. Nomura, K.; Ohta, H.; Ueda, K.; Hirano, M.; Hosono, *Science* **2003,** 300, 1269-1272.
2. Nomura, K.; Ohta, H.; Takagi, A.; Kamiya, T.; Hosono, H., et al. *Nature* **2004,** 432, 488-492
3. Cao, Q.; Hur, S. H.; Zhu, Z. T.; Sun, Y.; Rogers, J. A., *Adv. Mater.* **2006,** 18, 304-309.
4. Ju, S. Y.; Facchetti, A.; Xuan, Y.; Liu, J.; Ishikawa, F.; Ye, P. D.; Zhou, C. W.; Marks, T. J.; Janes, D. B., *Nat. Nanotechnol.* **2007,** 2, 378-384.
5. Durkop, T.; Getty, S. A.; Cobas, E.; Fuhrer, M. S., *Nano Lett.* **2004,** 4, 35-39f
6. Thostenson, E. T.; Ren, Z. F.; Chou, T. W., *Compos. Sci. Technol.* **2001,** 61, 1899-1912
7. Kang, S. J.; Kocabas, C.; Ozel, T.; Rogers, J. A. et al., *Nat. Nanotechnol.* **2007,** 2, 230-236
8. Liu, P.; Sun, Q.; Zhu, F.; Liu, K.; Jiang, K.; Li, Q.; Fan, S., *Nano Lett.* **2008,** 8, 647-651.
9. Rogers, J. A.; Bao, Z.; Drzaic, P., et al., *Proc. Natl. Acad. Sci. U. S. A.* **2001,** 98, 4835-4840

Mater. Res. Soc. Symp. Proc. Vol. 1144 © 2009 Materials Research Society 1144-LL07-08

Hydrothermal Synthesis and Photocatalytic Activity of Titanium Dioxide Nanotubes, Nanowires and Nanospheres

Jin Wang[1], Ming Li[1], Mingjia Zhi[1], Ayyakkannu Manivannan[2], Nianqiang Wu[1,*]

[1]Department of Mechanical and Aerospace Engineering, West Virginia University, Morgantown, WV 26506-6106, USA
[2]Department of Physics, West Virginia University, Morgantown, WV 26506, USA
*Corresponding author: Fax: +1-(304)-293-6689, E-mail: nick.wu@mail.wvu.edu

ABSTRACT

TiO_2 nanostructures with various morphologies and crystal structures were obtained by calcination of alkaline hydrothermal synthesized hydrogen titanate at different temperatures. The photocatalytic activities of the as-prepared samples were investigated by degradation of methyl orange aqueous solution under ultraviolet irradiation. The effects of the phase composition, crystallinity, surface area and shape of the nanostructures were evaluated. The results showed that high crystallinity, pure anatase phase and nanowire structure are favorable for the photocatalysis. The dependence of the photocatalytic activity on the surface area is not significant as usually expected.

INTRODUCTION

Titanium dioxide is widely used in photocatalytic and photoelectrochemical systems [1-2]. Recently, an alkaline hydrothermal method was demonstrated to be capable of scale-up production of one-dimensional (1-D) titanate nanostructures such as nanotubes and nanowires [3]. TiO_2 nanotubes have a high specific surface area and an open mesoporous morphology, which are expected to facilitate the adsorption of reactants on the active surface sites [4]. For TiO_2 nanowires, preliminary theoretical modeling and calculation work have showed that materials with such 1-D morphology may possess higher charge transport property as compared to nanospheres [5]

The photocatalytic and photoelectrochemical performances of the catalyst materials are governed by several factors such as surface area, particle size, pore volume and distribution, crystal structure and phase composition, surface and bulk defects, impurity species, morphological structure as well as surface hydroxyl group and etc. [6-8]. Given the conflicting results from the previous studies on the photocatalysis of TiO_2 materials, it remains unclear how these factors play their roles in the photocatalysis process. The inconsistency among the reported results mostly stems from the use of catalysts obtained from different synthesis routes, which leads to the difference in the specific surface properties of the products, such as defects concentration, crystallinity and etc. Thus the pivotal step toward better understanding the effects of the different parameters (particle size, phase composition and etc.) on the photocatalytic performance is to develop a synthesis route which is capable of producing TiO_2 materials with controllable and repeatable properties (crystal structure, morphology, surface area and etc). The alkaline hydrothermal synthesis method is capable of tailoring TiO_2 nanomaterials into different structures and shapes in a controlled fashion [9]. Thus the use of alkaline hydrothermal synthesized TiO_2 materials for the photocatalysis studies could minimize the negative impacts of the synthetic route variety on the photocatalytic performance.

In our experiments, a series of TiO_2 nanomaterials with different crystal structures (anatase, rutile, TiO_2-B) and shapes (nanowires, nanotubes, nanospheres) were synthesized by varying the hydrothermal temperature and the calcination temperature. Photocatalytic activities of these materials were evaluated by degrading methyl orange in an aqueous solution under irradiation of ultraviolet light.

EXPERIMENTAL

The TiO_2 nanostructures were synthesized by the hydrothermal process. Briefly, 1.2 g anatase titanium dioxide particles were added to the 80 ml 10 M NaOH aqueous solution. The mixture was vigorously stirred for 1 hour and then transferred to a 100 ml Teflon-lined stainless steel autoclave. The autoclave was sealed and then put into a preheated oven to perform hydrothermal treatment at different temperatures (120 °C and 200 °C respectively) for 24 hours to generate nanotubes and nanowires, respectively. White fluffy powder was obtained by hydrothermal processing. The white fluffy product was washed with deionized water and 1 M hydrochloric acid until the pH of the washing solution is less than 7. The washed products were then dried in an oven at 80 °C overnight. The as-washed samples were then calcinated in a ceramic boat inside a quartz tube furnace in the ambient environment. The temperature of the furnace was ramped at a constant rate (1 °C /min) to the calcination temperature and held for 4 hours. Three calcination temperatures (400 °C, 700 °C and 800 °C) were selected. After calcination, the samples were cooled to room temperature at a rate of 2 °C /min.

The crystal structure of samples was characterized by X-ray powder diffraction (XRD) (X'Pert Pro PW3040-Pro, Panalytical Inc.) with a Cu Kα radiation. The morphology was observed with a Hitachi S4700 field-emission scanning electron microscope (FE-SEM). The Brunauer-Emmett-Teller (BET) specific surface area was measured by the nitrogen adsorption using Accelerated Surface Area and Porosity System 2020 (Micromeritics).

The photocatalytic activity of the TiO_2 nanostructures was evaluated by measuring the decomposition of the aqueous solution of methyl orange. The photocatalysis experiments were carried out in a commercial photoreactor (LUZ-4V, Luzchem) which is equipped with the 148 W UVA lamp (Centered at 350 nm, LZC-UVA, Luzchem). A total of 10 mg TiO_2 catalysts were added to 10 ml solution of 20 mg/L methyl orange in a 10 ml polyethylene tube. Before irradiation, the suspensions were sonicated in the dark for 5 minutes. During the irradiation, the tubes were placed onto the carousel inside the photoreactor to ensure the even exposure of each tube to the UVA light. At the different irradiation time intervals, the tubes were unloaded and then centrifuged at 10,000 rpm for 1.5 hour to separate the supernate and the catalysts. The supernates were collected and analyzed by recording the characteristic absorption of methyl orange (464 nm) using a UV-Vis spectrometer (Shimadzu UV-2550). According to the calibration plot of the UV absorption as a function of the remaining methyl orange concentration, the efficiency of the methyl orange decomposition was calculated.

RESULTS AND DISCUSSION

Hydrothermal treatment of TiO_2 in the concentrated alkaline solution led to the formation of sodium titanate nanostructures. Nanotubes were formed at the synthesis temperature $T \leq 150$ °C while nanowires were obtained at T>150 °C. Further washing the as-synthesized sodium titanate products with aqueous acid solution facilitates the ion exchange between H^+ and Na^+, resulting in the hydrogen titanate. Calcination of the hydrogen titanate at higher temperature induces the loss of H_2O molecules from the layered hydrogen titanate structure and results in the formation of titanium dioxide. Various polymorphs of titanium dioxide can be obtained by varying the calcination temperature. Details of the phase transformation of hydrothermal synthesized titania nanowires and nanotubes can be found in our previous publication [9].

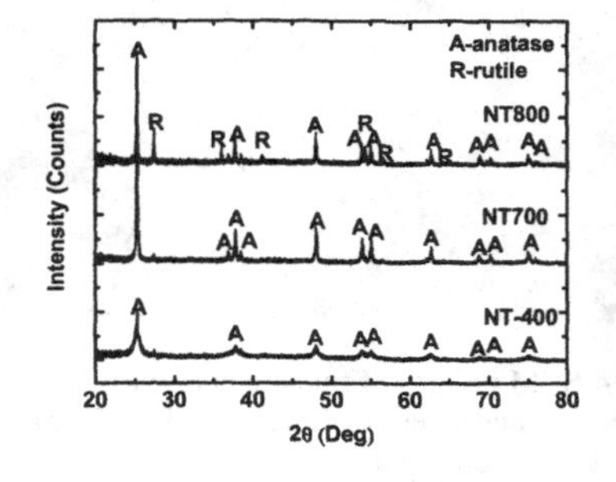

Fig. 1 XRD patterns of titanate nanotubes after heat treated at different temperatures for 4 hours

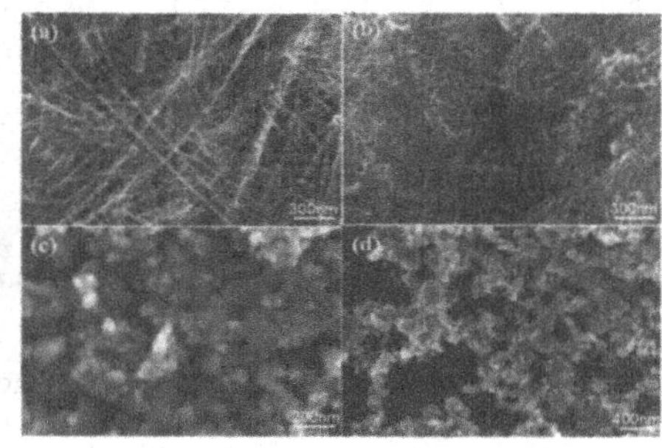

Fig. 2 SEM images of titanate nanotubes before (a) and after heat treated at 400 °C (b); 700 °C (c); 800 °C (d).

Figure 1 shows the XRD patterns of the hydrogen titanate nanotubes after calcination at 400 °C, 700 °C and 800 °C, respectively. Figure 2 shows the SEM images of the hydrogen titanate nanotubes before and after calcination at 400 °C, 700 °C and 800 °C, respectively. The hydrogen titanate nanotubes have an open-end and multi-wall structure with a diameter of around 15 nm and a length of hundreds of nanometers. Heat treatment of the hydrogen titanate nanotubes lead to the phase transformation to titania and degradation of tubular structure. It can be seen from the XRD and SEM results that three kinds of materials were derived from annealing of the hydrogen titanate nanotubes. Annealing at 400 °C (denoted as NT400) resulted in the anatase TiO_2 nanotubes with relatively poor crystallinity. Calcination at 700 °C (denoted as NT700) led to the anatase TiO_2 nanospheres with good crystallinity. Treatment at 800 °C (denoted as NT800) induced a mixture of nanospheres and microplates with the mixed anatase and rutile phases.

Figure 3 shows the XRD patterns of hydrogen titanate nanowires after calcination at 400 °C, 700 °C and 800 °C, respectively. Figure 4 shows the SEM images of the hydrogen titanate nanowires before and after calcination at 400 °C, 700 °C and 800 °C, respectively. The hydrogen titanate nanowires are 50 to hundreds of nanometers wide and up to tens of micrometers long. It is worth noting that TiO_2 (B) phase occurred in the nanowire after calcination at 400 °C (denoted as NW400). TiO_2 (B) is a polymorph of titanium dioxide composed of edge and corner sharing TiO_6 octahedrea, which has a slightly lower density than rutile, anatase or brookite and a relatively open structure with significant voids and continuous channels [10]. Calcination of the hydrothermally processed hydrogen titanate nanowires at 700 °C resulted in the anatase TiO_2 nanowires (denoted as NW700). Increasing the calcination temperature to 800 °C (denoted as

NW800) led to the formation of a mixture of TiO_2 nanospheres and nanorods with the mixed anatase and rutile phases.

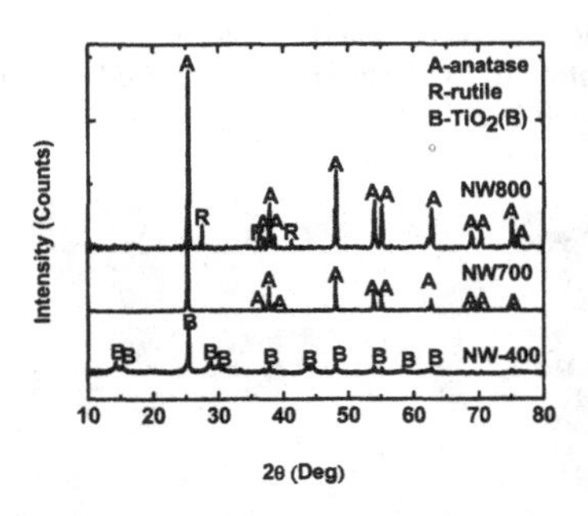

Fig. 3 XRD patterns of titanate nanowires after heat treated at different temperature for 4 hours

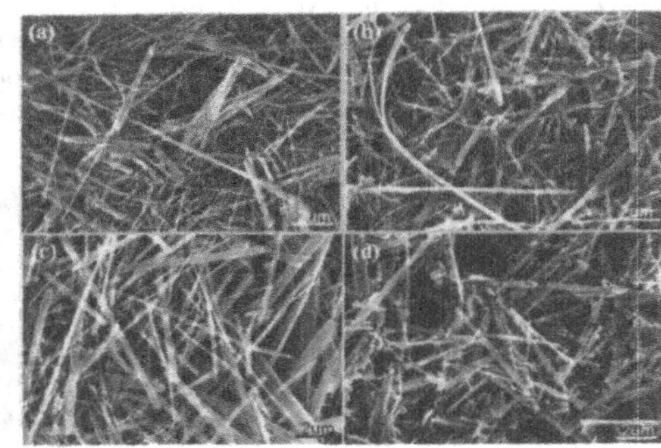

Fig. 4 SEM images of titanate nanowires before (a) and after heat treated at 400 °C (b); 700 °C (c); 800 °C (d).

The photocatalytic performances were evaluated in terms of the efficiency of degradation of methyl orange in an aqueous solution catalyzed by the synthesized materials. In the present work, the materials tested were derived from the same hydrothermal synthetic route with the same precursor. We only varied the calcination temperature to narrow the photocatalysis-controlling parameters down to the morphology, the surface area, the crystallinity and the phase composition. Figure 5 shows the apparent rate constants of methyl orange degradation by the as-synthesized catalysts. Among the NT-series samples, NT700 exhibited the highest photoactivity and NT400 showed the lowest activity. The photocatalytic activity of NW-series samples revealed the same rank order: NW700>NW800>NW400. The overall performance of a photocatalysis reaction by TiO_2 depends on two steps: (i). the separation of electrons and holes which involves two competitive processes—the generation of the electrons and holes by excitation and the recombination of the electrons and holes by recombination centers in the crystal lattice ; (ii). transfer of the electrons and holes to the substrate species pre-adsorbed on the catalyst surface. The BET surface area of NT400 was measured to be 219.86 m^2/g. With a large surface area and open-end hollow structure, it is supposed to allow more reactants to be adsorbed on the surface on the unit mass basis and to facilitate faster diffusion of the reactants and products. The BET surface area of NW700 was 20.83 m^2/g, which was 10 times lower than that of nanotubes. However, NW700 shows much higher photoactivity than NT400. This is ascribed to the fact that crystallinity plays a dominant role in the photocatalysis of methyl orange among the governing factors including crystallinity, surface area and particle size. Higher calcination temperature generally leads to higher photocatalytic activity, probably because high temperature annealing eliminates the thermal defects in the crystal lattice which could act as the recombination centers. The phase constituent also influences the photocatalytic efficiency. The pure anatase is the most preferred phase for photocatalysis application of both the nanotube and the nanowire catalysts. The mixed phase catalysts do not show enhanced photocatalytic activity.

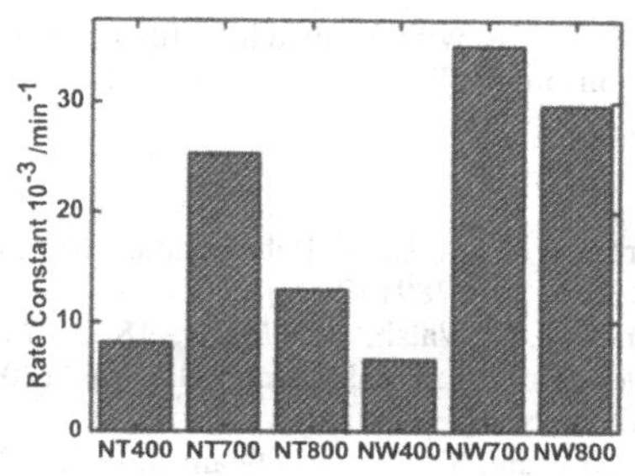

Fig. 5 Photocatalysis rate constant of titanate nanotubes and nanowires heat treated at different temperature under UVA irradiation

Compared to NT700 and NW700, the slight decrease in the photocatalytic activity in NT800 and NW800 could be due to the morphology change. Although TiO_2-B nanowires had been found to be prominent lithium intercalation materials [11], it shows the lowest photocatalytic activity among all the materials studied in our experiments. Comparing the NT-series with the NW-series, it is obviously seen that the nanowires showed much higher photoactivity than its nanotube and nanosphere counterparts regardless of the phase structure, even though nanowires have a much lower surface area. It has been demonstrated that the nanowire structure is a superior structure for charge transport which makes it an interest element in sensing applications [12]. The nanowire structure might provide continuous and directional channel for charge mobility, which greatly facilitates the separation of photo-generated electrons and holes. On the other hand, the relatively high crystallinity of the single-crystal nanowire also rendered a reduced recombination rate of charge carriers.

CONCLUSIONS

Various TiO_2 polymorphs (anatase, rutile, TiO_2-B) and morphologies (nanotubes, nanowires, nanosphers) were prepared by calcination of the hydrothermal synthesized hydrogen titanate materials. The photocatalytic activities of the as-prepared samples were investigated by degradation of methyl orange aqueous solution under UVA irradiation. It was found that the crystallinity plays a crucial role in the photocatalysis. Materials calcined at lower temperature showed inferior performances despite of their higher surface area. Pure anatase is the preferred phase for photocatalysis. TiO_2-B has the lowest activity in photocatalysis. The anatase nanowires showed better photocatalytic performance than their counterparts of nanotube or nanosphere with the same phase composition. It is inferred that the nanowire structure is favorable for generation of electron-hole pairs and suppression of the electron-hole recombination.

ACKNOWLEDGEMENTS

This work is financially supported by NSF grant (CBET-0834233) and West Virginia State Research Challenge Grant (EPS08-01). Some of the facilities and resources used in this work are supported by NSF grant (EPS 0554328) with the matching funds from the West Virginia University Research Corporation and the West Virginia EPSCoR Office.

REFERENCES

1. M. R. Hoffmann, S. T. Martin, W. Choi, D. W. Bahnemann, Chem. Rev., 95, 69 (1995).
2. X. Chen, S. S. Mao, Chem. Rev., 107, 2891 (2007).
3. D. V. Bavykin, J. M. Friedrich, F. C. Walsh, Adv. Mater., 18, 2807 (2006).
4. K. Zhu, N. R. Neale, A. Miedaner, A. J. Frank, Nano Lett., 7, 69 (2007).
5. J. P. Lewis, G-N. Tafen (private communication)
6. H. Kominami, S. Murakami, J. Kato, Y. Kera, B. Ohtani, J. Phys. Chem. B, 106, 10501 (2002).
7. S. S. Watson, D. Beydoun, J. A. Scott, R. Amal, Chem. Engr. J., 95, 213 (2003).
8. W. H. Leng, Z. Zhang, J. Q. Zhang, C. N. Cao, J. Phys. Chem. B, 109, 15008 (2005).
9. J. Wang, N. Q. Wu in *Nanotechnology Research Advances*, edited by X. Huang (Nova Science Publishers, Inc., New York, 2008), pp.241-260.
10. A. R. Armstrong, G. Armstrong, J. Canales, P. G. Bruce, Angew. Chem. Int. Ed., 43, 2286 (2004).
11. G. Armstrong, A. R. Armstrong, P. G. Bruce, P. Real, B. Scrosati, Adv. Mater., 18, 2597 (2006).
12. Y. Huang, X. Duan, Y. Cui, L. J. Lauhon, K. H. Kim, C. M. Lieber, Science, 194, 1313 (2001).

Mater. Res. Soc. Symp. Proc. Vol. 1144 © 2009 Materials Research Society 1144-LL07-09

Enhanced 1540 nm emission from Er-doped ZnO nanorod arrays via coupling with localized surface plasmon of Au island film

Jiang-Wei Lo, Chin-An Lin, and Jr-Hau He
Institute of Photonics and Optoelectronics, & Department of Electrical Engineering, National Taiwan University, Taipei, 10617 Taiwan (ROC)

ABSTRACT

Self-assembled nanorod arrays (NRAs) heterostructures that consist of a single-crystalline Er-doped ZnO NRAs grown on Au nanodot films have been synthesized by a chemical method and proposed as one of the promising optoelectronic materials since the Er intra-4f shell transition leads to 1540 nm emission for optoelectronic communication. The enhancement of 1540-nm emission of Er-doped ZnO NRAs via enhanced deep level emission of ZnO host resulted from local field enhancement effects of Au nanodot films, and subsequent energy transfer to Er^{3+} has been demonstrated. The microstructural analysis, electronic structure analysis, and photoluminescence characterizations have been performed to clarify the mechanism of enhanced 1540 nm emission. This paves the way to electrical pumping in a nano-system that forms NRAs of high-quality optical cavity.

INTRODUCTION

Er-doped semiconductors have been investigated intensively as one of the promising optoelectronic materials since the Er intra-4f shell transition leads to a 1540-nm emission, which lies in the minimum loss region of silica-based optical fibers. Every effort has been made for enhancing the Er-related emission of 1.54 mm by annealing,[1] oxygen doping,[2] and the use of wide-band-gap material[3] such as an amorphous Si,[4]or a SiC,[5] or a GaN.[6]

Zinc oxide has been considered as a promising candidate as a host material for Er doping because oxygen is a major element in the composition of ZnO with a wide band gap of about 3.3 eV. Er doped-ZnO films have proved reliable materials for light-emitting diodes, laser diodes, and optical amplifiers at 1540 nm in the waveguide structure as well as electrode materials for carrier injection because of their high electrical conductivity.[7] However, the solubility of Er in ZnO is quite low, leading to low luminescence efficiency.[8] Overcoming this limitation requires nanometer-scale design of the structure. Recently, it has been reported on successful optical activation of randomly-oriented ZnO nanowires by Er ion implantation or chemical method, [2,9] suggesting that one-dimensional ZnO nanostructures have potential as a new material platform for optical communication.

In addition, the surface plasmon (SP) has been found to enhance the light emitting efficiency by metal particles coating.[10] For example, Okamoto et al. enhanced significantly the photoluminescence (PL) intensity of the internal quantum efficiency of InGaN quantum wells by coating silver layer.[11] Visible light emission of ZnO thin films was greatly improved via coupling SP of Ag nanodots.[12]

In this work we investigated on self-assembled chemical growth of a dense nanorod array (NRA) heterostructures that consist of single-crystalline Er-doped ZnO NRAs grown on Au nanodot films/Si, providing the enhancement of the Er^{3+} luminescence intensity via coupling deep level emission of ZnO host with SP resonance, and subsequent energy transfer to Er^{3+}. The microstructural analysis, electronic structure analysis, and the luminescence mechanism are investigated. This opens the possibility of electrical pumping in a nano-system that forms NRAs of high-quality optical cavity naturally.

EXPERIMENT

The ZnO NRAs on the substrates were grown using hydrothermal process. We prepared the silicon (100) substrate coated with 5-, 10-, 20-, and 40-nm-thick Au nanodot films by e-gun evaporator. A thin film of zinc acetate was spin-coated on the bare Si and Au nanodot films/Si substrates for ten times with a solution containing 5 mM zinc acetate dihydrate added to the ethanol. ZnO films of 5-10 nm thickness were obtained after annealing at 300 ℃ in air for 20 min. ZnO NRAs were grown in aqueous solution containing zinc nitrate hexahydrate (10 mM) and ammonia solution at 95 ℃ for 2 hr. The reactants then cooled to room temperature and the substrate was washed with ethanol to remove impurities. After drying, the as-grown ZnO NRAs were then spin-coated with erbium chloride hexahydrate in ethanol at 2000 rpm for 10 sec with a holding time of 10 sec. The as-coated samples were then annealed at 800℃ for 4 hours in air ambient.

The structures of Er-doped ZnO NRs on Au-coated Si were characterized using the x-ray diffraction (XRD) with the x-ray source of a Cu K_α line and scanning electron microscope (SEM). The electronic structures of Er-dope ZnO NRAs were characterized using x-ray photoelectron spectroscopy (XPS). The UV/visible and IR PL measurements of ZnO NR arrays with Au nanodot films were performed using 325nm He-Cd laser.

DISCUSSION

Hydrothermal method is adapted to grow ZnO NRAs since it is easy to grow ZnO nanorods perpendicularly to the surface of arbitrary substrates regardless of their lattice match or surface chemistry.[13,14] Before ZnO NRAs were grown from the textured ZnO seeds in aqueous solution, several thicknesses of Au with identical thermal treatment were selected to prepare suitable Au nanostructures on the Si substrates for coupling with the ZnO NRAs. The surface morphologies of Au nanodot films with different thicknesses ranging from 5 to 40 nm on the Si substrates were shown in the up-left insets of Fig. 1(b)-(e). As the thickness of Au thin film is increased, the size of Au nanodot structures is expected to become large with the irregular shape. It has been known that the size and shape of metal nanodots determine the resonance energy of a generated SP. The surface morphologies of the ZnO NRAs grown directly on bare Si substrates and Si substrates with different thicknesses of Au nanodot films are shown in the down-left insets of Fig. 1(a)-(e). Note that the growth conditions of ZnO NRAs are identical in all samples. Compared to the ZnO NRA growth on Si substrates, the heterogeneous nucleation for ZnO NRA growth on Au/Si substrates can be much easier since stable nucleus is already present with the aid of Au nanodot films. As a result, the diameter of ZnO NRAs increases with the thickness of the Au films under identical conditions. The as-synthesize ZnO NRAs were then spin-coated uniformly with ~3.5-nm-thick $ErCl_3$.[15] Figure 1(a)-(e) shows the SEM images of the ZnO NRAs with $ErCl_3$ on the different thicknesses of Au/Si substrates after annealing at 800℃ for 4 hours in air ambient.

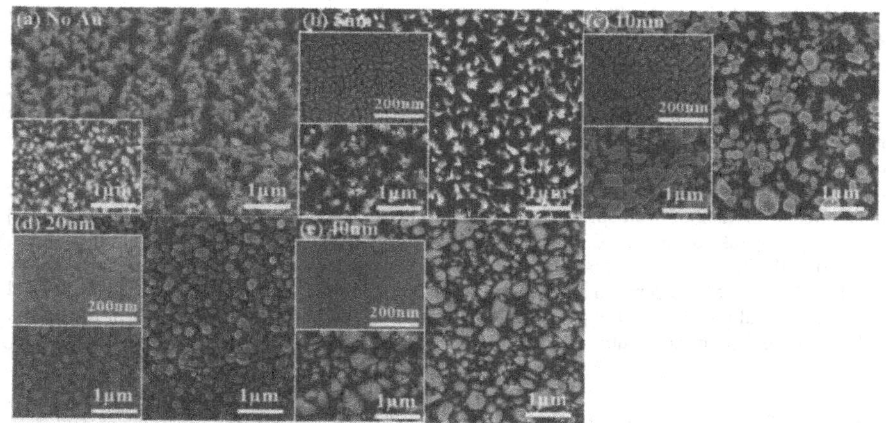

Figure 1. The SEM images of the ZnO NRAs with $ErCl_3$ on the different thicknesses of Au/Si substrates. The surface morphologies of Au nanodot films with different thicknesses ranging from 5 to 40 nm on the Si substrates were shown in the up-left insets of Fig. 1(b)-(e). The surface morphologies of the ZnO NRAs grown directly on bare Si substrates and Si substrates with different thicknesses of Au nanodot films are shown in the down-left insets of Fig. 1(a)-(e).

The structures of Er-doped ZnO NR arrays were investigated by the x-ray diffraction (XRD) spectra. Figure 2 shows that the XRD spectra of the Er-doped ZnO NRAs samples on Au/Si substrates. All the diffraction peaks can be indexed to wurtzite ZnO structure. It is worth mentioning that no Er_2O_3 diffraction peaks were observed in the XRD patterns of all samples, indicating the absence of Er_2O_3 either on the ZnO surface or in the ZnO structures.

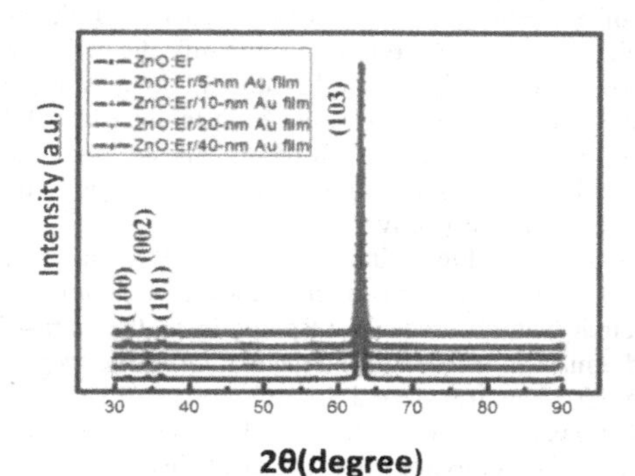

Figure 2. The XRD spectra of the Er-doped ZnO NRAs on the different thickness of Au/Si substrates.

The chemical bonding structures of the films were analyzed for the Zn_{2p3}, O_{1s}, and Er_{4d} orbitals using XPS. Figure 3 shows the XPS spectra from : (a) O_{1s}, (b) $Zn_{2p3/2}$, and (c) $Er_{4d5/2}$ core levels in the Er-doped ZnO NRAs on Au nanodot films/Si substrates. The surface was cleaned by Ar^+ ion etching prior to XPS measurement. A chemical shift of the binding energy (BE) was calibrated by C_{1s} signal from the adsorbed surface as being 285.0 eV. The O 1s signal presented a line shape due to the contributions at BE=531 eV, ascribed to Zn-O-Zn moieties. Correspondingly, the BE of the Zn 2p3/2 was very close to 1022.2 eV, referring to zinc(II) oxide and displayed negligible variations under various conditions.[16] The Er 4d signal presented a complex shape, related to coupling phenomena between the 4d hole and the lanthanide partially filled 4f shell. In particular, whatever the processing conditions, the shape and position of the most intense 4d5/2 spin-orbit split component at 168.5 eV were typical for Er(III) species.[16,17] Moreover, A peak located at 529 eV,[15] attributed to Er_2O_3, was not observed in all samples. Therefore, the XPS data are consistent with the results obtained from the XRD spectra revealing that Er atoms were incorporated into wurtzite ZnO lattice rather than formed cubic Er_2O_3 crystal with 4-hr annealing. However, a peak at 529 eV can be found in the samples with 5-hr annealing, indicating the formation of cubic Er_2O_3.[15]

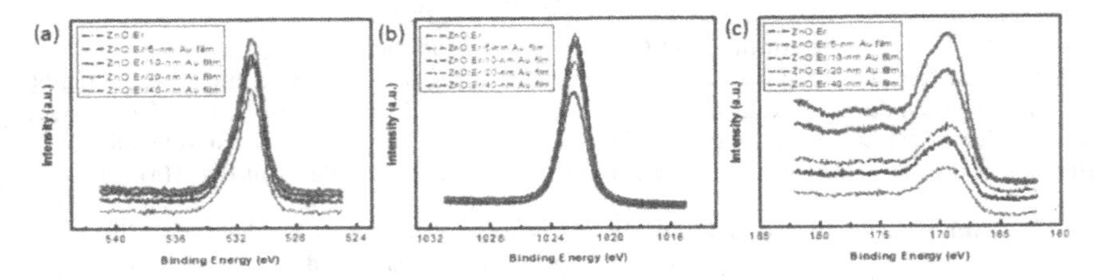

Figure 3. The XPS spectra from : (a) O_{1s}, (b) $Zn_{2p3/2}$, and (c) $Er_{4d5/2}$ core levels in the Er-doped ZnO NRAs on the different thicknesses of Au/Si substrates.

Since the number of photoelectron of an element depends on the atomic concentration of that element in the samples, XPS is used to not only identify the elements but also quantify the chemical composition. We can check the possibility that the increasing ratio of 520-nm/382-nm emission is induced by the deficiency of O with different thicknesses of Au nanodot films. The compositions of Zn, O and Er are invariable with the thickness of Au nanodot films because of the ratio of these integrated intensities of about unity. The average compositions are 51.5%, 45.8%, and 2.7% for O, Zn, and Er, respectively.

The IR PL spectra, as shown in Fig. 4, indicate that IR PL intensity of Er-doped ZnO NRAs increases with thickness of Au nanodot films under 325-nm excitation. A main peak centered at 1540 nm in all PL spectras was observed corresponding to the intra-4f shell transition. The mechanism of Er-related emission is through the generation of electron-hole pairs in ZnO host, and the absorbed energy is then transferred to Er^{3+} ions.[7,9,18] Afterward, the 1540 nm emission results from the relaxation of Er^{3+} ion from the first excited state ($^4I_{13/2}$) to the ground state ($^4I_{15/2}$). The 1540 nm photoemission also provides the evidence that Er is successfully doped into ZnO lattice by a wet chemical reaction on various Au nanodot film/Si substrates. In addition, fine structures have been observed in the Er-related PL spectra. These fine structure also comes from the intra-4f shell transition of the Er^{3+} ions, and is not a result of the PL interference within the Er-doped ZnO NRAs because there is no change in the PL feature for changing the incident angle of the excitation laser beam. The shape of the PL spectra in IR region remains unchanged on various Au nanodot film/Si substrates. This indicates that Er-doped ZnO NRA growth on different Au nanodot film/Si substrates brings same placement of Zn and O atoms around Er since the spectrum shape concerns local structure around Er.[19]

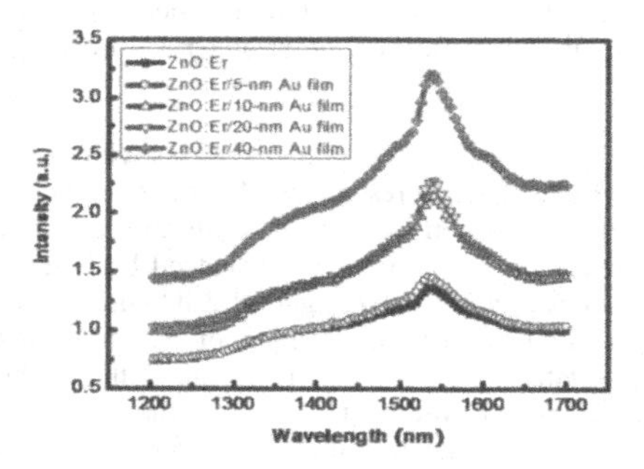

Figure 4. The IR photoluminescence spectra of the Er-doped ZnO NRAs on the different thicknesses of Au/Si substrates.

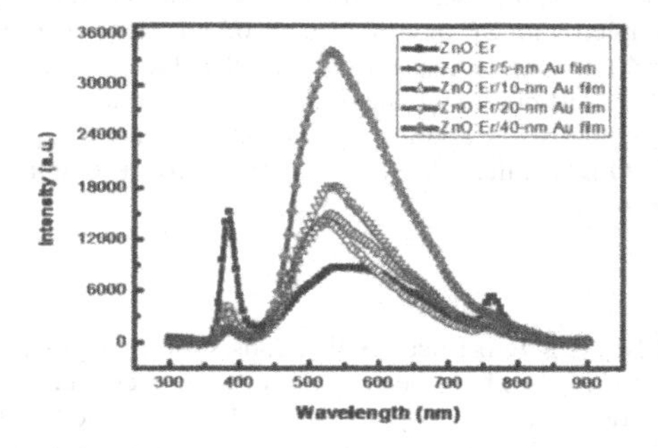

Figure 5. The UV/visible photoluminescence spectra of the Er-doped ZnO NRAs on the different thicknesses of Au/Si substrates.

According to XPS measurements, the invariable ratio shows the amount of Er incorporated into ZnO in all samples did not account for enhancing the 1540-nm emission. To investigate the origin of enhanced IR PL, UV/visible PL measurements have been performed. Figure 5 shows that the UV/visible PL spectra of all samples exhibited a NBE emission at 382 nm and a DEL at 520 nm, which attributed to high density of native defects such as oxygen vacancies and the Zn vacancies.[14,20,21] It was found that the 520 nm emission increases with the thickness of Au nanodot films. The ratio of 520 nm emission to 382 nm emission of Er-doped ZnO NRAs can be increased by over 36 times. We can exclude the possibility that the giant enhanced 520-nm emission is induced by deteriorated crystal quality, such as oxygen deficiency according to XPS measurements. It has been investigated that the enhancement of DLE emission of ZnO is via coupling with SP of Au.[22] The enhancement of DLE emission strongly depends on the size of Au nanodots. It has been known that the energy of the SP resonance depends on the size and the shape of these metal particles.[23] The larger Au nanodots lead to the slight redshift of the reflection maximum with increasing intensity.[24,25] The broad spectra are due to the irregular size of the Au nanodots. In addition, this difference of enhanced ratio of visible emission to UV emission disproves the possible attribution of the stronger PL intensity of ZnO NRAs to the metal reflection.

Consequently, the density of 1540-nm emission increases with the thickness of Au nanodot films due to the enhanced surface plasmon resonance of Au nanoparticles. While the sizes of Au nanodots increase, the surface plasmon resonance scattering emission dominates over the absorption process in visible region, resulting in the enhanced DLE emission (~520nm) of ZnO NRAs. The deep level states existing in the band gap of ZnO could serve as the energy transfer media for the Er ion excitation.[7,9] The wavelength of 520 nm corresponds to the energy transition between the $^2H_{11/2}$ state and the ground $^4I_{15/2}$ state of the 4f shell of Er^{3+}, opening the more possibility of energy transfer towards Er^{3+}; i.e. Er ions in the ground states resonantly absorb the energy from the emission related to deep level of ZnO host. The enhancement of the emission of the 1540 nm emission from $^4I_{13/2}$ to $^4I_{15/2}$ was observed.
The non-linear dependence of IR PL intensity on the intensity of the deep-level emission showed that the presence of the ZnO native defects shall have two opposite effects on the 1540-nm luminescence. On the one hand, they serve as the resonant states that can be pumped by the laser excitations and then coupled with SP modes. The absorbed energy can be transferred to the Er ions and lead to the 1540-nm luminescence. On the other hand, the native defect-related deep level states also affect the efficiency of the 1540-nm emission, as consideringthe ZnO NBE emission, deep level green emission, and the Er emissions that are competing with each other. In addition, some of the ZnO native defects serve as nonradiative recombination centers, leading to the quench of the luminescence.

CONCLUSIONS

Self-assembled NRAs heterostructures that consist of a single-crystalline Er-doped ZnO NRAs grown on Au nanodot films have been synthesized by a chemical method. The deep level emission of ZnO host plays a role in energy transfer to Er^{3+} ions due to resonant absorption. The enhancement of 1540-nm emission of Er-doped ZnO nanorod arrays are via enhanced deep level emission resulted from local field enhancement effects of Au nanodot films, and subsequent resonant absorption to Er^{3+} ions.

ACKNOWLEDGMENTS

The research was supported by the National Science Council Grant No. NSC 96-2112-M-002-038-MY3.

REFERENCES

1. Y. Ishikawa, M. Okamoto, S. Tanaka, D. Nezaki, N. Shibata, *J. Mater. Res*, **20** No. 9, Sep (2005).
2. K. Takahei, A. Taguchi, *J. Appl. Phys.* **74**, 1979 (1993).
3. P. N. Favennec, H. L'Haridon, M. Salvi, D. Moutonnet, T. Le Guillou, *Electron. Lett.* **25**, 718 (1989).
4. M. S. Bresler, O. B. Gusev, V. Kh. Kudoyarova, A. N. Kuznetsov, P. E. Pak, E. I. Terukov, I. N. Yassievich, B. P. Zakharchenya, W. Fuhs, A. Sturm, *Appl. Phys. Lett.* **67**, 3599 (1995).
5. A. J. Steckl, J. Devkajan, W. J. Choyke, R. P. Devaty, M. Yoganathan, S. W. Novak, *Journal of Electronic Materials* **25**, 869 (1996).
6. A. J. Steckl, J. Heinkenfeld, D. S. Lee, M. J. Gartner, C. C. Baker, Y. Wong, R. Jones, *IEEE J. Sel. Top. Quantum Electron.* **8**, 749 (2002).
7. S. Komuro, T. Katsumata, T. Morikawa, X. Zhao, H. Isshiki, Y. Aoyagi, *Appl. Phys. Lett.* **76**, 3935 (2000).
8. E. Sonder, R.A. Zhur, R.E. Valiga, *J. Appl. Phys.* **64**, p. 1140 (1998).
9. J. Wang, M. J. Zhou, S. K. Hark, Q. Lia, D. Tang, M. W. Chu, C. H. Chen, *Appl. Phys. Lett.* **89**, 221917 (2006).
10. M. Fukushima, N. Managaki, M. Fujii, H. Yanagi, S. Hayashi, *J. Appl. Phys.* **98**, 024316 (2005).
11. K. Okamoto, I. Niki, A. Shvartser, Y. Narukawa, T. Mukai, A. Scherer, *Nat. Mater* **3**, p. 601 (2004).
12. P. Cheng, D. Li, Z. Yuan, P. Chen, D. Yang, *Appl. Phys. Lett.* **92**, p. 041119 (2008).
13. M. S. Arnold, P. Avouris, Z. W. Pan, Z. L. Wang, *J. Phys. Chem. B* **107**, 659 (2003).
14. K. Vanheusden, W. L. Warren, C. H. Seager, D. R. Tallant, J. A. Voigt, B. E. Gnade, *J. Appl. Phy.* **79** , p. 7983 (1996).
15. W. C. Yang, C. W. Wang, J. C. Wang, Y. C. Chang, H. C. Hsu, Tzer-En Nee, L. J. Chen, J. H. He, *J. Nanosci. Nanotechnol*, **8**, 3363–3368 (2008).
16. L. Armelao, D. Barreca, G. Bottaro, A. Gasparotto, D. Leonarduzzi, C. Maragno, E. Tondello, C. Sada, *J. Vac. Sci. Technol.* **24**, Issue 5, pp. 1941-1947 (2006).
17. W. C. Lang, B. D. Padalia, L. M. Watson, D. J. Fabian, P. R. Norris, *Faraday Discuss. Chem. Soc.*, **60**, 37 – 43 (1975).
18. A. Polman, *J. Appl. Phys.* **82**, 1 (1997).
19. M. Ishii, S. Komuro, T. Morikawa, Y. Aoyagi, *J. Appl. Phys.* **89**, 3679 (2001).
20. K. Vanheusden, C. H. Seager, W. L. Warren, D. R. Tallant, J. A. Voigt, *Appl. Phys. Lett.* **68**, p. 403 (1996).
21. S. A. Studenikin, N. Golego, M. Cocivera, *J. Appl. Phy.* **84**, p.2287 (1998).
22. C. A. Lin and J. H. He, in preparation.
23. P. C. Andersen, K. L. Rowlen, *Applied Spectroscopy* **56**, p. 124A (2002).
24. J. R. Lakowicz, *Analytical Biochemistry* **337**, p. 171 (2005).
25. R. Gans, *Ann. Physik* **37**, p. 811 (1912).

Mater. Res. Soc. Symp. Proc. Vol. 1144 © 2009 Materials Research Society 1144-LL13-13

Effects of Laser Ablation on Growth of ZnO/ZnS/ZnO Multilayer Structured Nanorods by Chemical Vapor Deposition

Takashi Hirate, Hiroaki Koisikawa, Makoto Yugi, Takuya Kumada, Yuki Matsuzawa and Tomomasa Satoh
Department of Electronics and Informatics Frontiers
Faculty of Engineering, Kanagawa University
3-27 Rokkakubashi, Kanagawa-ku, Yokohama, 221-8686, Japan

ABSTRACT

We study effects of laser ablation of Mn on growth of ZnS:Mn layer on ZnO nanorods grown by chemical vapor deposition cooperated with laser ablation of Mn and on growth of ZnO(Mn) layer on the ZnS:Mn layer. It is concluded that the laser ablation of Mn performed during growth of ZnS:Mn layer and ZnO(Mn) layer promotes the growth of each layer. And we fabricated ZnO nanorods double-coated with ZnS:Mn layer and ZnO(Mn) layer.

INTRODUCTION

Zinc oxide (ZnO), a wide-direct-bandgap (3.37 eV) semiconductor, is an attractive II-VI compound semiconductor material for various optoelectronic devices. Recently, growth of various nanostructures of ZnO such as nanorod, nanobelt, nanowall, etc. has been reported [1-11], and ZnO has been considered as a promising material for nanodevices. We have studied on application of ZnO nanorods to electroluminescent devices. In this paper we report on fabrication of ZnO nanorods double-coated with ZnS:Mn layer and ZnO(Mn) layer to apply ZnO nanorods to DC-operated electroluminescent devices, cooperating laser ablation of Mn during growth of ZnS and ZnO by chemical vapor deposition (CVD).

EXPERIMENTAL DETAILS

ZnO nanorods coated with ZnS:Mn and ZnO(Mn) layers are sequentially grown on an n-Si (111) substrate by CVD method cooperated with laser ablation that is developed by us. The growth equipment is similar that reported before [1, 8]. This is basically a low-pressure thermal CVD system. The precursors are metal Zn vapor and O_2 or H_2S to synthesize ZnO or H_2S respectively. Zn vapor is introduced into a deposition chamber through orifice whose diameter is 1 mm. N_2 is used as carrier gas to transport Zn vapor from Zn vaporization chamber to

Table 1. Growth conditions for ZnO:Mn/ZnS:Mn/ZnO nanorods growth.

	1st. ZnO	2nd. ZnS:Mn	3rd. ZnO:Mn
Zn Evaporation Temperature [C]		585	
Flow Rate of N2 for Zn [SCCM]		7	
Flow Rate of O_2 or H_2S [SCCM]	0.8 (O_2)	1.0 (H_2S)	0.8 (O_2)
Growth Temperature [C]	500	400	500
Growth Pressure [Pa]		13	
Growth Time [Min]	15	25	15
Laser Power [W]	0.6	0.8	0.6

deposition chamber. O_2 or H_2S is also introduced into deposition chamber. A Mn (99.99%) target (10 mm diameter, 10 mm height) is set in deposition chamber. A pulsed Nd:YAG laser beam (wavelength =1.064 μm, pulse width = 10 ns, repetition frequency = 10 Hz) irradiates the Mn target surface through the quartz window. The growth conditions are listed in Table 1.

Firstly, ZnO nanorods are grown on an n-Si (111) substrate with laser ablation of Mn in initial 3 min of growth. Then, after changing of growth temperature to 400 C in 5 min, ZnS:Mn layer is grown. The laser ablation of Mn is performed during the growth. Thirdly, after returning of growth temperature to 500 C in 5 min, ZnO(Mn) layer is grown. The laser ablation of Mn is also performed during the growth.

The surface morphology was examined using a field-emission scanning electron microscope (FE-SEM, HITACHI S-4000). The crystal structure was analyzed using an X-ray diffractometer (XRD, RIGAKU RU-200, Cu-Kα1). The photoluminescence spectrum was measured in a wavelength range between 300 nm and 900 nm under He-Cd laser (325 nm) excitation.

RESULTS AND DISCUSSION

Figure 1 shows the cross-sectional FE-SEM image of one sample of ZnO nanorods firstly grown on n-Si (111) substrate. The height of nanorods is 2400nm and the diameter is about 44 nm. Although the diameters of the first ZnO nanorods do not so vary between each growth, the height and direction slightly vary in each growth even under same growth conditions. When the Mn laser ablation in initial 3 min of growth is not performed, randomly oriented nanorods are grown and vertically aligned nanorods are not grown.

Figure 2 shows cross-sectional FE-SEM images of nanorods coated with second (a) ZnS:Mn layer with laser ablation of Mn and (b) ZnS layer without laser ablation. The first ZnO nanorod of Figure 2(a) is as shown in Figure 1. A diameter of ZnS:Mn-coated nanorods with

Figure 1. Cross-sectional FE-SEM image of one sample of ZnO nanorods firstly grown on n-Si (111) substrate.

laser ablation is about 80 nm in average. The second ZnS:Mn layer grows with about 18 nm thickness over the whole surface of the first ZnO nanorods, but the surface of the second ZnS:Mn layer is slightly rough although the surface of the first ZnO nanorods is smooth. When the laser ablation is not performed, second ZnS layer grows only a little as shown in Figure 2(b). Although the FE-SEM image of the first ZnO nanorods of this sample is not shown here, the diameter is almost same as that shown in figure 1.

Figure 3 shows XRD patterns of samples

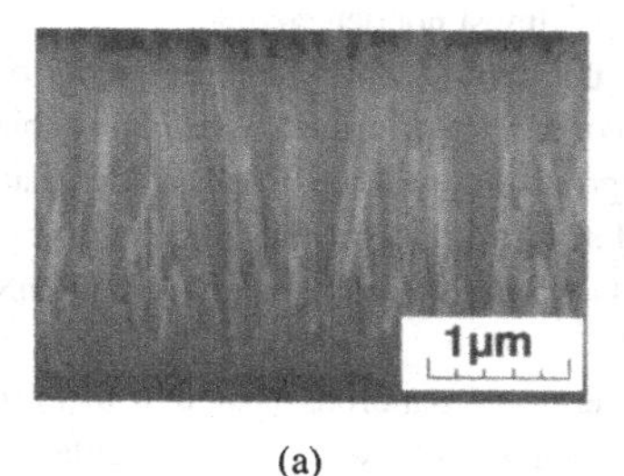

(a)

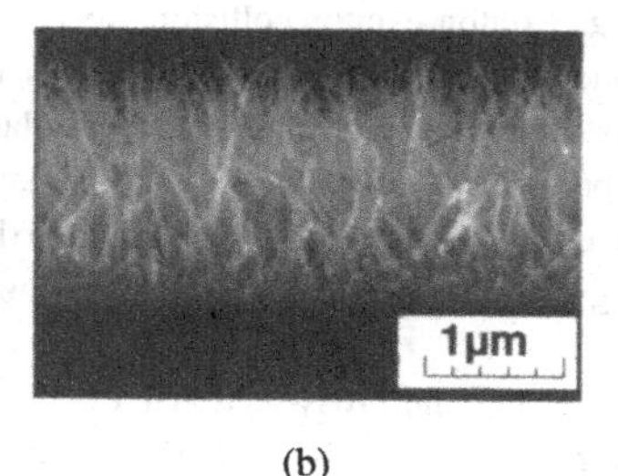

(b)

Figure 2. Cross-sectional FE-SEM images of ZnO nanorods coated with second (a) ZnS:Mn layer with laser ablation of Mn and (b) ZnS layer without laser ablation.

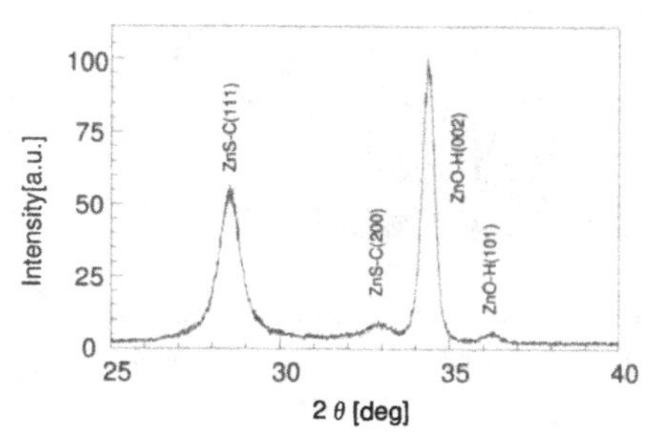

Figure 3. XRD patterns of nanorods coated with second ZnS:Mn layer shown in figure 2(a).

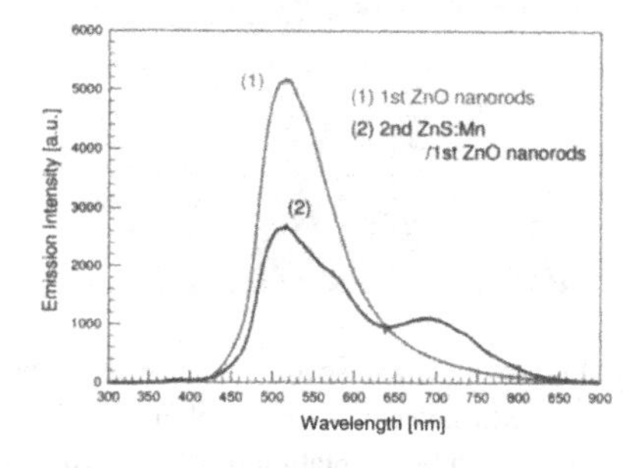

Figure 4. PL spectra of (1) ZnO nanorods shown in figure 1 and (2) ZnO nanorods coated with second ZnS:Mn layer shown in figure 2(a).

shown in figure 2(a). The diffraction lines of ZnO-H (hexagonal ZnO) (002) and ZnO-H (101) are from first ZnO nanorods, and ZnS-C (cubic ZnS) (111) and ZnS-C (200) from second ZnS:Mn layer. It is concluded that the first ZnO nanorods grow with c-axis orientation comparing with J.C.P.D.S. standard (No. 36-1451) data of hexagonal (wurtzite) ZnO powder, and that the second ZnS:Mn layer grows uniformly with polycrystalline cubic crystal structure without any directional growth comparing with J.C.P.D.S. standard (N0. 5-566) data of cubic (zinc-blende) ZnS. The XRD results are consisted with the SEM images.

Figure 4 shows PL spectra of samples shown in figure 1 and figure 2(a). The green band emission around 530 nm was generally explained by the radiative recombination of a photo-generated hole with the electron in a slightly ionized oxygen vacancy in ZnO [12]. The emissions around 580 nm and 680 nm are from Mn^{2+} in ZnS [13, 14]. The UV emission around 380 nm related to a near band-edge transition of ZnO, namely, the recombination of free excitons through exciton-exciton collision process [15] is almost not detected.

It is concluded from experimental results that the second ZnS:Mn layer grows on all surfaces of first ZnO nanorods when the laser ablation of Mn is performed. When laser ablation of Mn is not performed, however, second ZnS layer grows only a little. We consider that any species produced by laser ablation of Mn and arrived at the surface of the first ZnO nanorods promotes the growth of second ZnS:Mn layer suppressing directionality of growth of ZnS:Mn layer.

Figure 5 shows the cross-sectional FE-SEM images of nanorods coated with third (a) ZnO(Mn) layer with laser ablation of Mn and (b) ZnO layer without laser ablation on

Figure 5. Cross-sectional FE-SEM images of ZnO nanorods coated with second ZnS:Mn layer with laser ablation of Mn and further coated with third (a) ZnO(Mn) layer with laser ablation of Mn and (b) ZnO layer without laser ablation.

ZnS:Mn/ZnO nanorods. When laser ablation is performed, third ZnO(Mn) layer grows on all surfaces of ZnS:Mn/ZnO nanorods, being the mean diameter 110 nm. When laser ablation is not

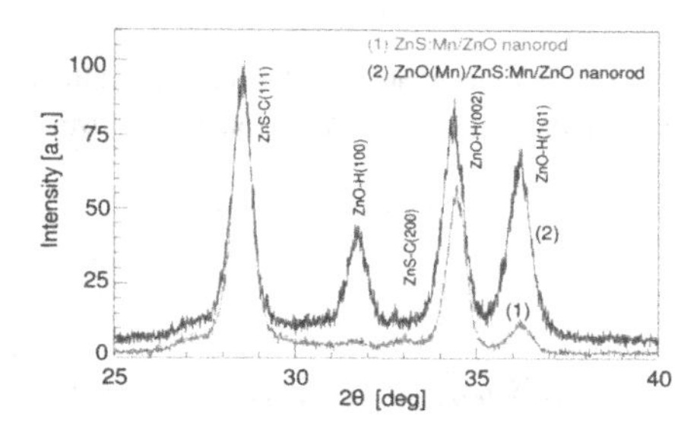

Figure 6. XRD patterns of (1) ZnS:Mn / ZnO nanorods and (2) ZnO(Mn) / ZnS:Mn / ZnO nanorods.

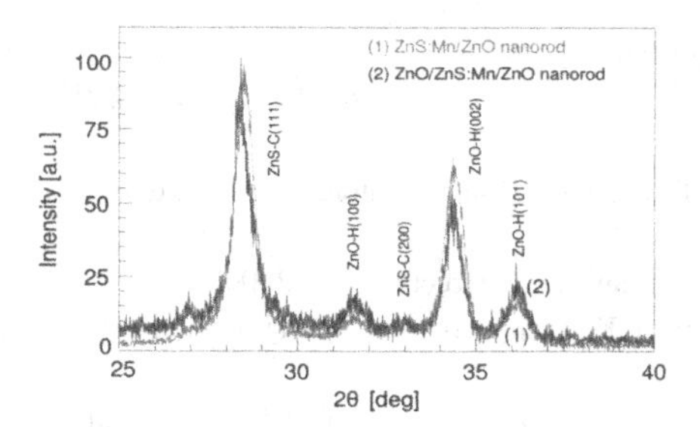

Figure 7. XRD patterns of (1) ZnS:Mn / ZnO nanorods and (2) ZnO / ZnS:Mn / ZnO nanorods.

performed, however, third ZnO does not grow as a layer but grows as short (200 nm) whiskers. The number density of whiskers is high at the top area of ZnS:Mn/ZnO nanorods as shown in figure 5(a).

Pattern (2) in figure 6 shows XRD pattern of the sample shown in figure 5(a) and pattern (1) in the figure is that of sample before growth of third ZnO(Mn) layer for reference. The intensities of ZnO-H (100) and ZnO-H (101) diffractions increase considerably, although that of ZnO-H (002) does not so increase. This means that the third ZnO(Mn) layer does not grow with c-axis directionality as in the case of growth of first ZnO nanorods but grows on all surfaces of ZnS:Mn/ZnO nanorods without any directionality.

Pattern (2) in figure 7 shows XRD pattern of the sample shown in figure 5(b). Pattern (1) is that of sample before growth of third ZnO layer for reference. In this case, XRD patterns does not so change by growth of third ZnO whiskers, being due to relatively small volume of third ZnO whiskers and to random distribution of axis direction of whiskers.

It is concluded that the third ZnO(Mn) layer grows on all surfaces of ZnS:Mn/ZnO nanorods when the laser ablation of Mn is performed. When laser ablation of Mn is not performed, the third ZnO grows as whiskers with short length. We consider that any species produced by laser ablation of Mn also suppresses a directional growth of ZnO and results in growth as layer.

It is revealed experimentally that laser ablation of Mn has drastic effects on growth of second ZnS:Mn layer and third ZnO(Mn) layer. We consider that layered structure of ZnO(Mn) / ZnS(Mn) / ZnO nanorods is promising for development of a new luminescent device including ZnO nanorods.

CONCLUSIONS

We developed a technology to fabricate ZnO nanorods coated with polycrystalline ZnS:Mn layer and polycrystalline ZnO(Mn) layer by incorporation of laser ablation of Mn during CVD of ZnO and ZnS.

REFERENCES

1. T. Hirate, N. Takei, and T. Satoh, *Proceedings of the 2002 International Conference on the Science and Technology of Emissive Displays and Lighting*, 2002, p.81.
2. S. X. Mao and M. Zhao, Appl. Phys. Lett. 83 (2003) 993.
3. Y. J. Xing, Z. H. Xi, Z. Q. Xue, X. D. Zhang, J. H. Song, R. M. Wang, J. Xu, Y. Song, S. L. Zhang, and D. P. Yu, . Appl. Phys. Lett. 83 (2003) 1689.
4. M. Y.an, H. T. Zhang, E. J. Widjaja, and R. P. H. Chang, Appl. Phys. Lett. 83 (2003) 5240.
5. J. H. Choy, E. S. Jang, J. H. Won, J. H. Chung, D. J. Jang, and Y. W. Kim, Appl. Phys. Lett. 84 (2004) 287.
6. P.X.Gao and Z.L.Wang, Appl. Phys. Lett. 84 (2004) 2883.
7. B. P. Zhang, K. Wakatsuki, N. T. Binh, Y. Segawa, and N. Usami, J. Appl. Phys. 96, (2004) 340.
8. T. Hirate, S. Sasaki, W. Li, H. Miyashita, T. Kimpara, and T. Satoh, Thin Solid Films, 487(2005)35.
9. H. Miyashita, T. Satoh, and T. Hirate, Superlattices and Microstructures, 39(2006)67.
10. R. C. Wang, C. P. Liu, J. L. Huanga, S. J. Chen, Y.K. Tseng, and S.-C. Kung, Appl. Phys. Lett. 87 (2005) 013110.
11. X. Zhang, Y. Zhang, J. Xu, Z. Wang, X. Chen, D. Yu, P. Zhang, H. Qi, and Y. Tian, Appl. Phys. Lett. 87 (2005) 013111.
12. K. Vanheusden, W. L. Warren, C. H. Seager, D. K. Tallant, J. A. Voigt, B. E. Gnade, J. Appl. Phys., 79 (1996) 7983.
13. M. D. Bhise, M. Katiyar, and A. H. Kitai, J. Appl. Phys., 67 (1990) 1492.
14. J. Benoit, P. Benalloul, A. Geoffroy, N. Balbo, C. Barthou, J. P. Denis, and B. Blanzat, Phys. Status Solidi A, 83 (1984) 709.
15 Y. C. Kong, D. P. Yu, B. Zhang, W. Fang, S. Q. Feng, Appl. Phys. Lett., 78 (2001) 407.

Mater. Res. Soc. Symp. Proc. Vol. 1144 © 2009 Materials Research Society 1144-LL14-05

Growth and Characterization of p-n Junction Core-Shell GaAs Nanowires on Carbon Nanotube Composite Films

Parsian K. Mohseni[1], Gregor Lawson[2], Alex Adronov[2], and Ray R. LaPierre[1]

[1] Centre for Emerging Device Technologies, Department of Engineering Physics, McMaster University, Hamilton, Ontario, L87 4L7, Canada

[2] Department of Chemistry, McMaster University, Hamilton, Ontario, L87 4L7, Canada

ABSTRACT

Thin films composed of poly(ethylene imine)-functionalized single-walled carbon nanotubes (CNTs) were formed through a vacuum filtration process and decorated with Au nanoparticles, roughly 40 nm in diameter. The Au nanoparticles, on the surface of the CNT fabric, accommodated the growth of GaAs nanowires (NWs), according to the vapour-liquid-solid (VLS) mechanism, in a gas-source molecular beam epitaxy (GS-MBE) system. Structural analysis indicated that the nanowires, up to 2.5 μm in length, were not preferentially oriented at specific angles with respect to the substrate surface. The NWs grew in the energetically favored [0001] direction of the wurtzite lattice while stacking faults, characterized as zincblende insertions, were observed along their lengths. Micro-photoluminescence spectroscopy demonstrated bulk-type optical behaviour. Current-voltage behaviour of the core-shell pn-junction heterostructured NWs exhibited asymmetric rectification. Thus, the potential for the incorporation of such hybrid NW/CNT architectures into an emerging class of flexible opto-electronic devices is demonstrated.

INTRODUCTION

To date, semiconductor nanowires (NWs) and carbon nanotubes (CNTs) have been the focus of intensive research within the realm of nanometer scale science and engineering [1-2]. Of particular interest are CNT composite films that have been demonstrated as promising materials for flexible optoelectronics applications due to their impressive optical transparency and electrical conductivity [3-4]. Also, semiconductor nanowires, which can be grown on dissimilar surfaces [5] via the Au-assisted vapour-liquid-solid (VLS) mechanism, have been successfully incorporated as the active medium into novel NW-based light-emitting diodes and photovoltaic cells [6-7]. Here, we establish for the first time, hybrid III-V semiconductor NW/CNT-film architectures assembled through gas-source molecular beam epitaxy (GS-MBE). The nature of the growth surface is discussed in terms of the site specific nucleation of each nanowire. Characterization of the structural properties, optical quality, and electrical behaviour of the NWs grown on the CNT fabrics is presented. This work represents the initial step towards a novel class of flexible photonic devices, based on hybrid NW/CNT architectures.

EXPERIMENTAL DETAILS

Single-walled CNTs were covalently functionalized with poly(ethylene imine) (PEI), and formed into 1-2 μm thick films via a vacuum filtration technique [8]. Au nanoparticles, roughly 40 nm in diameter, were reduced on the CNT film surface by submerging the film in an $HAuCl_4$ solution for 60 seconds. The Au decorated films were subjected to a 10 minute annealing sequence in nitrogen ambient at a temperature of 550 °C. Upon annealing, the films were loaded in a GS-MBE system. Nanowire growth was initiated and maintained at 550 °C, by introducing a flux of Ga monomers and As_2 dimers to the surface of the Au decorated CNT films, at a V/III flux ratio of 1.5. A nominal GaAs film growth rate of 1.0 μm/hour was maintained for a period of 30 minutes, at which point NW growth was terminated by closing the shutter of the group III effusion cell. Two NW groups were investigated in this study: Group A NWs were undoped (unintentional in-situ n-doping is unavoidable), while group B NWs were initially doped with Te for 15 minutes, then doped with Be for a final 15 minutes, both at concentrations of 10^{18} cm^{-3}, as calibrated using previous GaAs films grown on GaAs (100) substrates. Thus, group B NWs inherently contain a p-n (shell-core) heterostructure, while group A NWs do not.

A JEOL JSM-7000 scanning electron microscope (SEM), a Carl Zeiss SMT NVision 40 dual-beam focused ion beam (FIB) microscope, a Philips CM12 transmission electron microscope (TEM), and a JEOL 2010F high-resolution transmission electron microscope (HR-TEM) were employed for the various structural characterization experiments presented here. Micro-photoluminescence (μ-PL) characterization was performed using a continuous flow helium cryostat at 7 K, using a laser emission centered at 532 nm and excitation power of 80 μW.

DISCUSSION

A planar SEM image of a typical sample of Au nanoparticles on a CNT film, after an annealing process, is shown in Figure 1. The diameter distribution of the nanoparticles was centered at roughly 40 nm. The average nanoparticle density was determined to be approximately 50 μm^{-2}, with particles being separated by an average pitch of nearly 90 nm. These conditions allow for dense NW growth while limiting the merging of neighboring NWs due to sidewall deposition and shadowing effects.

Figure 1. Post-annealing SEM images showing distribution of Au nanoparticles on CNT film.

An SEM image of the as-grown group A NW sample is shown in Figure 2. Here, densely grown NWs can be observed on the surface of the flexible CNT fabric. The NW full-width at half-length diameter was measured as an average of roughly 100 nm, while NW lengths ranged between 1 to 2.5 μm. Typical NWs were of a tapered morphology and orientated in a variety of angles with respect to the substrate. The wide range in orientation angles is attributed to the lack of a periodically structured and epitaxially guiding substrate, as would be the case when using conventional single phase semiconductor substrates. Nonetheless, the NWs tend to grow in the thermodynamically preferential [0001] direction of the wurtzite crystal lattices, as observed in a TEM image of a single group A NW, shown in Figure 3. The selective area electron diffraction (SAED) pattern shown in the inset of Figure 3, confirms the aforementioned crystal structure and growth direction. Here, stacking faults are visible towards the NW base, which are characterized as zincblende lateral insertions formed under high group III supersaturation conditions [9]. The highly tapered morphology is attributed to sidewall-diffusion limited deposition, concurrent with VLS growth [5].

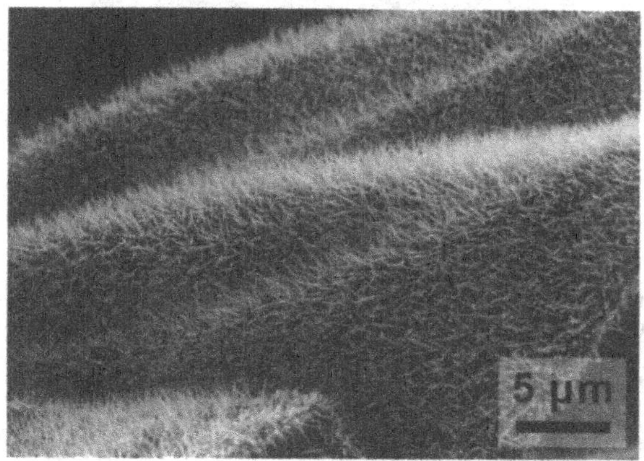

Figure 2. SEM image of as-grown group A NW sample. The flexibility of the film, containing densely packed and randomly oriented NWs, is clearly observed.

Figure 3. TEM image of single group A NW with tapered morphology. The SAED pattern shown in the inset confirms the growth direction along the c axis of the wurtzite crystal.

Of considerable importance is the nature of the growth surface in this particular study. As each Au nanoparticle initially lies upon a unique location, spanning in width across several randomly oriented single-walled CNTs, it is unclear weather the growth was epitaxially instigated. However, TEM analysis of lamellae removed from the as-grown samples, via FIB, indicated that the initial nucleation of GaAs indeed occurred at the Au/CNT interface. Figure 4 shows a TEM image of one such lamella where the base of a GaAs NW, a 2-dimensional film (deposited simultaneously with NW growth), and the CNT substrate can at once be examined. Here, stacking faults spanning the length of the visible NW, and protruding from within the planar GaAs region, are observed. The observation of stacking faults within the GaAs film, while localized to the extent of the NW width, demonstrates that the initial GaAs island nucleation leading to the growth of a single NW occurred at the Au/CNT interface.

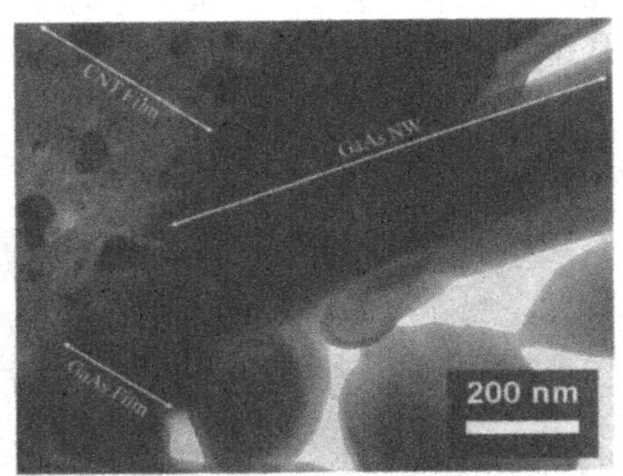

Figure 4. TEM image of the interfaces formed between a single GaAs NW, a planar GaAs film, and the CNT substrate from an as-grown sample. The observation of localized stacking faults near the NW base, and present throughout the GaAs films, indicates that NW growth was initiated at the surface of the CNTs.

Low temperature micro-photoluminescence characterization of single NWs showed a single peak centered at 1.51 eV, with a linewidth of approximately12.5 meV, as indicated by the spectrum shown in Figure 5. Temperature dependent studies indicated a peak energy red-shift with increasing temperature. This trend is plotted in the inset of the Figure 5 together with the theoretical temperature dependent bandgap of bulk GaAs, according to the Varshni relation. Thus, based on this agreement, the single NW μ-PL can be attributed to band-to-band recombination transitions within the bulk-like undoped GaAs NW. This demonstrates the high crystallinity and notable optical quality of the NWs, allowing for the potential incorporation into photonics devices.

Finally, we turn our attention to the electrical behaviour of the hybrid NW/CNT structures. To allow for electron transport through the NWs to the CNT film, while avoiding a short-circuit pathway to the GaAs film, devices were processed and contacted with Ti/Pt/Au, using the as-grown group A and B samples, in a manner described elsewhere [10]. Figure 6 shows the measured current-voltage (I-V) behaviour of a purified CNT film (squares), a typical ensemble of contacted group A NWs (circles), and a typical ensemble of contacted group B NWs (triangles). It is seen that the CNT films itself exhibits conductive behaviour, with a resistivity of roughly 6.8 × 10^{-3} Ω·cm. This value is competitive with commercial ITO films [4]. The group A NWs (undoped) exhibited a high resistivity of nearly 5900 Ω·cm, while group B (n-type core, p-type shell) NWs exhibited asymmetrically rectifying I-V behaviour, demonstrating the diode-type characteristic of the p-n junctions within the NWs. The potential, therefore, exists for the incorporation of similar hybrid architectures within flexible solar cell and light-emitters.

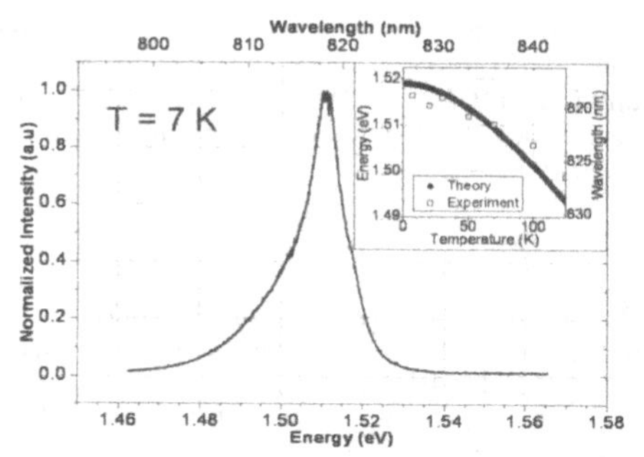

Figure 5. μ-PL spectrum from single undoped GaAs NW, indicative of bulk-like band-to-band recombination transitions. The inset plots the experimental temperature dependence of the PL peak together with the theoretical temperature dependent bulk GaAs bandgap.

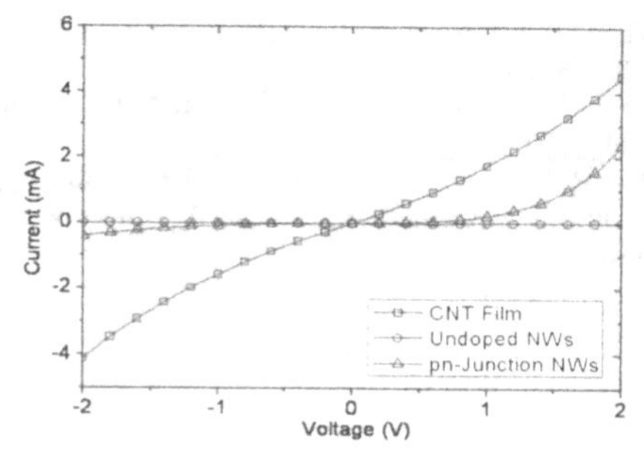

Figure 6. I-V characteristics of purified CNT film (squares), group A NWs (circles), and group B NWs (triangles). The diode-type asymmetric rectification evident from the group B NWs, while absent in the CNT film and group A NWs, demonstrates functional pn-junction core-shell heterostructured NWs.

CONCLUSIONS

We have presented the first integration of thin CNT films and GaAs NWs, grown according to the VLS mechanism via GS-MBE. The nanowires were shown to grow on the flexible surface of the CNT films and resemble, in structure, homoepitaxially grown NWs under otherwise similar growth conditions. The observations of strong PL signals and diode-type behaviour are promising for the incorporation of these hybrid CNT/NW architectures into flexible optoelectronics device applications.

ACKNOWLEDGMENTS

We would like to acknowledge the financial support of the Natural Sciences and Engineering Research Council (NSERC) of Canada, the Canadian Foundation for Innovation (CFI), the Ontario Innovation Trust (OIT), the Premier's Research Excellence Award, and the Ontario Centres of Excellence (OCE). Also, we are grateful to Brad Robinson for MBE growth and Christophe Couteau for μ-PL assistance.

REFERENCES

1. Y. Li, F. Qian, J. Xiang, and C. M. Lieber, *Materials Today* **9**, 18 (2006).
2. P. Avouris, M. Freitag, and V. Perebeinos, *Nat. Photonics* **2**, 341 (2008).
3. G. Gruner, *J. Mater. Chem.* **16**, 3533 (2006).
4. N. Saran, K. Parikh, D. S. Suh, E. Munoz, H. Kolla, and S. K. Manohar, *J. Am. Chem. Soc.* **126**, 4462 (2004).
5. P. K. Mohseni, C. Maunders, G. A. Botton, and R. R. LaPierre, *Nanotechnology* **18**, 445304 (2007).
6. M. S. Gudiksen, L. J. Lauhon, J. Wang, D. C. Smith, and C. M Lieber, *Nature* **415**, 617 (2002).
7. B. Z. Tian, X. L. Xeng, T. J. Kempa, Y. Fang, N. F. Yu, G. H. Yu, J. L. Huang, and C. M. Lieber, *Nature* **449**, 885 (2007) .
8. G. Lawson, F. Gonzaga, J. Huang, G. de Silveira, M. A. Brook, and A. Adronov, *J. Mater. Chem.* **18**, 1694 (2008).
9. P. K. Mohseni and R. R. LaPierre, *Nanotechnology* **19**, (2008) (in press).
10. P. K. Mohseni, G. Lawson, C. Couteau, G. Weihs, A. Adronov, and R. R. LaPierre, *Nano Lett.* **8**, (2008) (in press)

Mater. Res. Soc. Symp. Proc. Vol. 1144 © 2009 Materials Research Society 1144-LL16-04

Luminescence characterization of InGaN/GaN vertical heterostructures grown on GaN nanocolumns

Robert Armitage
Panasonic Electric Works Co., Ltd.
1048 Kadoma
Osaka, 571-8686 Japan

ABSTRACT

The photoluminescence of MBE-grown InGaN/GaN vertical heterostructures on *c*-axis oriented GaN nanocolumns is investigated. Nanocolumnar InGaN heterostructures exhibit luminescence efficiencies greater than 20% for peak emission wavelengths as long as 540 nm. Compared to otherwise identical InGaN samples with larger median column diameters, the luminescence is blue-shifted and exhibits reduced efficiency for diameters less than about 50 nm. Growth of InGaN on GaN columns with a broad distribution of diameters results in broad-band photo-luminescence that appears white to the eye and has efficiency as high as 23%.

INTRODUCTION

GaN and its alloys with InN and AlN are widely used in visible light-emitting diodes (LEDs) and other devices. Compared to conventional planar epilayers, synthesis of these materials in the form of nanocolumnar crystals offers some advantages including low densities of extended defects [1] and partial relaxation of strain in lattice-mismatched heterojunctions via the free surfaces of the columns [2,3]. The latter effect could potentially be exploited to extend the operation of InGaN LEDs to yellow and red wavelengths, which are difficult to realize with planar epilayers due to the 11% lattice mismatch between GaN and InN. In this work we study the photoluminescence of InGaN/GaN heterostructures on top of *c*-axis oriented GaN nano-columns grown spontaneously by molecular-beam epitaxy (MBE) on Si substrates.

EXPERIMENTS

Growth was performed in an Eiko EL-10A MBE machine with activated nitrogen supplied by an Oxford MPD-21 rf plasma source with a 276-hole PBN aperture plate and Ga and In evaporated from effusion cells. The base pressure is in the 10^{-10} torr range. Prior to growth the n-type Si(111) substrates were heated at 1050°C for 20 min. The rf source was operated at 450W with N_2 pressure of 4.5×10^{-5} torr. In some cases a 100-200 nm GaN buffer layer was deposited at 500°C before the GaN nanocolumns growth. Nanocolumns were grown at a substrate temperature of 850-920°C with a Ga cell temperature of 950-1000°C. The temperature was then lowered to 680-720°C for the growth of a thin (<20 nm) InGaN layer with the In cell at 850°C and Ga cell at 900°C. A 20-30 nm GaN cap layer was then grown at the same substrate temperature. All substrate temperatures refer to a thermocouple not directly in contact with the substrate, the reading of which is roughly 100°C higher than the true substrate temperature.

Nanocolumn morphologies were observed by field-emission scanning electron microscopy (SEM) with an acceleration voltage of 5 kV without any conductive coatings applied to the samples. Photoluminescence (PL) was excited with a HeCd laser with a maximum output of 50 mW and the luminescence was dispersed by a dual-grating monochromator and detected by a photomultiplier. The temperature was controlled by a closed-cycle He cryostat with a heating element. The PL spectra were not corrected for the spectral sensitivity of the optical system.

RESULTS AND DISCUSSION

Synthesis of GaN nanocolumns by MBE growth on Si(111)

The spontaneous growth of *c*-axis oriented GaN nanocolumns on Si(111) by MBE using high substrate temperatures and N-rich conditions has been studied by several groups [1,4-8]. We follow a similar approach and obtain results that agree qualitatively with those previously reported. The column diameters range from 20-150 nm and the height (400-800 nm) is controlled by the growth time. Each sample exhibits a distribution of column diameters rather than a single diameter. The columns overlap [8] for densities higher than 1×10^{10} cm^{-2} and median diameters greater than about 60 nm. Using GaN buffer layers [6] allows for lower column densities than growth directly on Si, but there is a broader distribution of column diameters and some tilting between the column axes and the Si [111] axis occurs, especially for buffer layers thicker than 100 nm. Examples of several representative column morphologies are shown in Fig. 1.

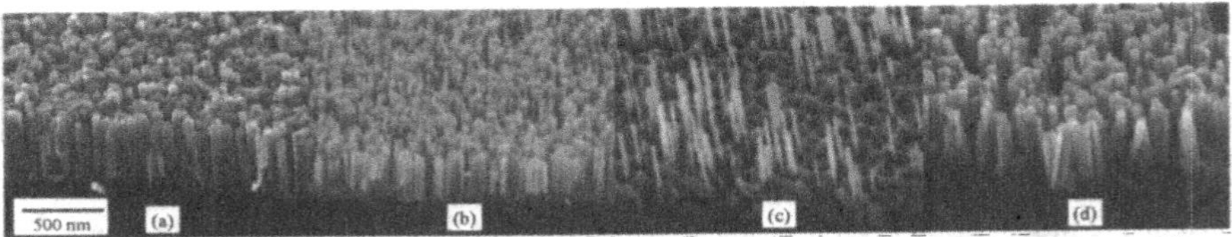

Fig. 1. SEM images of representative GaN nanocolumn morphologies: a) median diameter 65 nm, density 1×10^{10} cm^{-2}, directly on Si; b) 35 nm, 2×10^{10} cm^{-2}, directly on Si; c) 30 nm, 1×10^{9} cm^{-2}, on GaN buffer layer; d) 80 nm, 4×10^{9} cm^{-2}, on GaN buffer layer. Samples tilted 45°.

Characterization of InGaN photoluminescence with fixed underlying column morphologies

Compared to metalorganic chemical vapor deposition (MOCVD), MBE has generally proved not to be an effective technique for the growth of planar InGaN layers with efficient luminescence [9]. However, we have observed relatively high luminescence efficiencies in InGaN grown by MBE on GaN nanocolumns. Figure 2 (a) shows temperature-dependent PL spectra for a GaN cap (20 nm)/InGaN (9 nm) structure grown on GaN nanocolumns with median diameter of 80 nm. It is noteworthy that the InGaN PL peak position red-shifts monotonically with increasing temperature and does not show the "S-shaped" trend usually associated with locally inhomogeneous InGaN layers [10].

The ratio of the integrated PL intensity at room temperature to that at low temperature is taken as a measure of the PL efficiency (internal quantum efficiency) of InGaN layers [10]. The 295K/4K intensity ratio in Fig. 2 (a) is 22%, which is twice the highest PL efficiency reported for planar InGaN grown by MBE [9]. In fact, the InGaN nanocolumns PL efficiency of 22% rivals the efficiency of the best conventional planar InGaN/GaN heterostructures grown by any technique, provided that one compares samples with the same peak emission wavelength (530 nm). As shown in Fig.2 (b), the PL efficiency of conventional planar InGaN/GaN heterostructures decreases for emission wavelengths above 500 nm due to increasing mismatch strain as the In content increases [11]. In contrast, the PL efficiency of InGaN grown on GaN nanocolumns is essentially constant as the peak wavelength increases from 480 to 550 nm. Note that the only differences in growth conditions for the samples shown in Fig. 3 are slightly different InGaN growth temperatures used to modify the In contents in the layers. The trend shown in Fig. 2 (b) suggests that nanocolumns may indeed be useful for extending the spectral range of nitride LEDs to longer wavelengths.

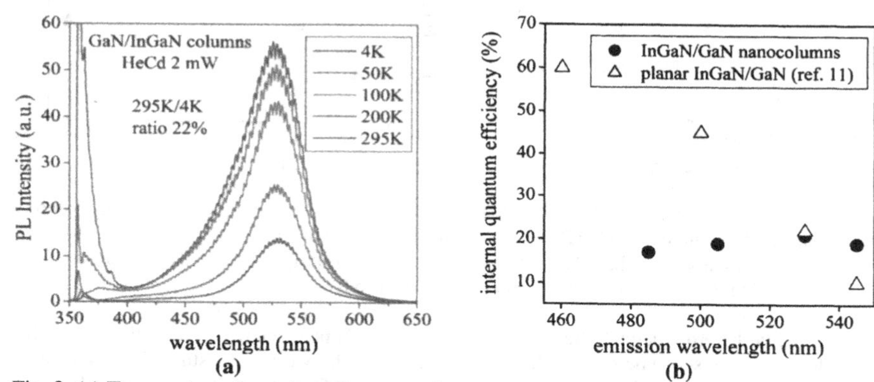

Fig. 2. (a) Temperature-dependent PL spectra for a sample with GaN column median diameter 80 nm, density 6×10^9 cm^{-2}, height 800 nm, InGaN thickness 9 nm, and 20 nm GaN cap. (b) PL efficiency as a function of peak emission wavelength for InGaN/GaN nanocolumns and conventional planar InGaN/GaN heterostructures grown on sapphire by MOCVD (ref. 11).

A reduction of the quantum-confined Stark effect (QCSE) due to relaxation of strain (and piezoelectric polarization) in nanocolumnar InGaN heterostructures could be an explanation for their relatively efficient PL despite high In contents. The QCSE can be investigated by changing the carrier concentration in an InGaN quantum well (QW) via changes in the excitation laser power. A blue-shift with increasing excitation power suggests the presence of the QCSE. In principle, filling of states associated with locally In-rich regions in InGaN may also cause a blue-shift with increasing excitation power. However, the presence of In-rich localized states is also associated with an "S-shaped" dependence of the InGaN PL peak position with increasing temperature, which we do not observe in any nanocolumn samples. Therefore we believe the blue-shift with increasing excitation power in InGaN nanocolumns is mainly due to the QCSE.

Excitation-power dependent PL data for a sample with emission peaking at 497 nm at room temperature are shown in Fig. 3 (a). The peak shifts 56 meV as the laser power is increased from 0.2 to 20 mW, which suggests that the QCSE is present to at least some extent in the sample. We observe smaller peak shifts with excitation power in samples with longer peak wavelengths; a shift is hardly detectable in a sample with room-temperature emission peaking at 530 nm. This seemingly contradictory behavior is tentatively attributed to higher unintentional donor concentrations in more In-rich samples grown at lower temperatures. These unintentional donors may screen the piezoelectric field, resulting in a reduced QCSE despite higher strain.

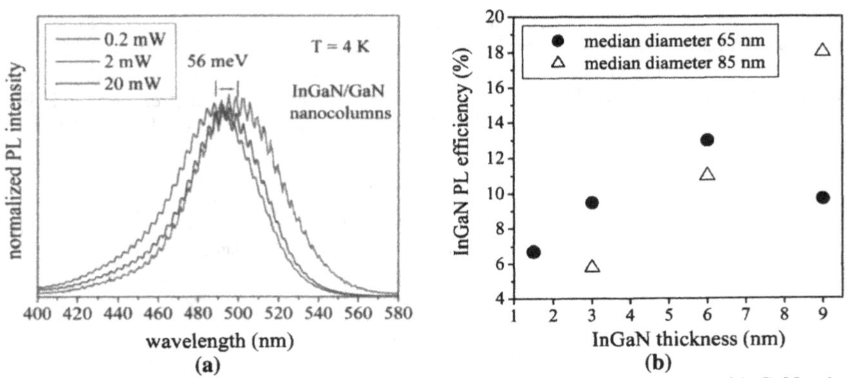

Fig. 3. (a) Shift of InGaN PL peak with increasing excitation power for sample with GaN column median diameter 85 nm, density 7×10^9 cm^{-2}, height 800 nm, InGaN thickness 9 nm, and 20 nm GaN cap measured at 4K. (b) InGaN PL efficiency as a function of InGaN thickness for two sample series with the InGaN thickness systematically varied.

In planar heterostructures the QCSE causes the InGaN peak energy to strongly vary with well width in sample sets where the well width is systematically varied. We studied two sets of samples where the InGaN thickness was systematically varied from 1.5 to 9 nm, but the PL peak position did not show a clear dependence on the thickness. The lack of a clear trend is not unexpected since the strain in nanocolumnar QWs is predicted to decrease with increasing well thickness [3]. The effect of increased well thickness may be offset by strain relaxation and reduced polarization charge in thicker wells. An interesting observation from the InGaN thickness series is that higher PL efficiencies are found in thicker InGaN layers, as shown in Fig. 3 (b). This may be related to more efficient strain relaxation in thicker layers, as discussed above.

Influence of column diameters on InGaN PL for fixed InGaN growth conditions

Figure 4 compares room-temperature PL spectra for two samples with identical growth conditions of the InGaN/GaN heterostructure but different morphologies of the underlying nano-columns. The thickness of the InGaN layer is approximately 3 nm. The peak PL peak for sample A (median diameter 35 nm) is 443 nm while that for sample B (median diameter 65 nm) is 490 nm, a difference much larger than the unintentional run-to-run variation (±5 nm). The PL efficiency is also lower in sample A (4%) than sample B (11%). We have compared four pairs of such samples, and in all four pairs the sample with smaller median column diameter (less than about 50 nm) showed a blue-shift of the InGaN PL peak and a decrease in InGaN PL efficiency.

The higher emission energy in Sample A cannot be attributed to lateral confinement in a quantum disk since the column diameter of 35 nm is much larger than the exciton (or hole) Bohr radius in InGaN. One factor likely contributing to the blue-shift is a diminished QCSE (reduced piezoelectric polarization) due to more efficient relaxation of heterostructure strain in smaller-diameter columns [3]. Another factor which may possibly contribute to the large blue-shift is the formation of In-rich rings around the peripheries of larger columns, which may occur to a lesser extent for smaller-diameter columns. The larger strain at the column center and larger strain variation along the radial direction provide a stronger driving force for In segregation to the periphery in larger-diameter columns [3]. Since recombination occurs preferentially in In-rich regions, material with a less homogenous In composition exhibits red-shifted PL emission for the same average composition [10].

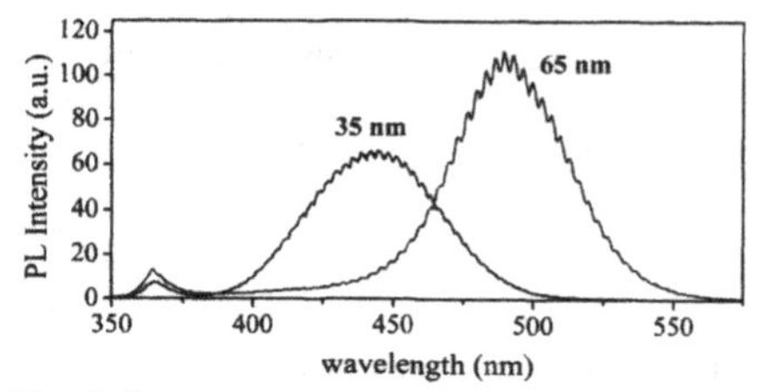

Fig. 4. Room-temperature PL spectra for nominally identical 3 nm-thick InGaN QWs grown on GaN columns with different median diameters (35 nm and 65 nm).

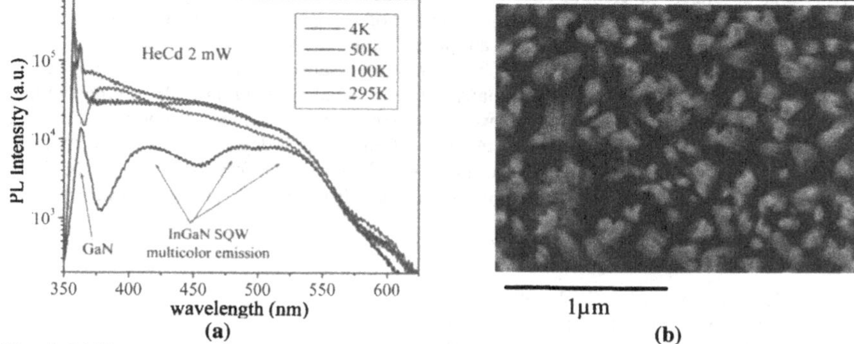

Fig. 5. (a) Temperature dependence of "white" PL spectra from InGaN grown on GaN columns with broad diameter distribution. (b) Bird's eye view SEM image of the sample morphology.

Two factors are considered as possible causes of the lower InGaN PL efficiency in samples with smaller column diameters (less than about 50 nm). One factor is the decrease in the surface barrier height as the column diameter decreases below the critical value at which the surface depletion regions originating from opposite sidewalls of the column overlap [7]. As the height of the potential barrier which separates electrons from holes trapped at the surface decreases, surface recombination becomes more significant. A second factor which may possibly reduce the InGaN PL efficiency in smaller diameter columns is surface confinement or spatially separated lateral confinement of electrons and holes due to non-uniform strain along the radial direction of the column [3]. The lateral confinement behavior can differ depending on the column diameter and would be further complicated in the case of non-uniform In distributions. Detailed calculations beyond the scope of this work are needed to investigate this possibility.

The change of the InGaN PL peak position depending on column diameter shown above explains why the PL spectra from nanocolumns with a distribution of diameters are broad compared to those of planar InGaN epilayers of similar In compositions. Indeed, an independent study of InGaN/GaN nanocolumns revealed that the PL spectra of individual columns are much narrower than those of column ensembles [12]. For several samples with single InGaN layers grown on GaN columns with especially broad diameter distributions, we observed very broad PL emission (Fig. 5) perceived by the eye as white light. A PL efficiency as high as 23% was observed in such "white" emitting samples, similar to the highest efficiency observed in "green" samples. InGaN/GaN nanocolumn assemblies with broad diameter distributions can potentially be used to produce white light from single LED chips without phosphor conversion.

CONCLUSIONS

InGaN grown on GaN nanocolumns exhibits PL efficiencies greater than 20%, significantly higher than those of planar InGaN layers grown by MBE. For peak wavelengths of 530 nm and beyond, the PL efficiencies of nanocolumns rival those of conventional planar InGaN/GaN structures grown by MOCVD. Although the nanocolumnar InGaN PL efficiency is relatively high at longer wavelengths, power-dependent PL measurements suggest the QCSE is present to at least some extent. Determining the degree of strain relaxation in nanocolumns and the corresponding reduction in the QCSE is a topic for future work.

The InGaN PL peak position and PL efficiency were shown to strongly depend on the column diameter in experiments with nominally identical InGaN growth on GaN columns with different median diameters. These findings were discussed in terms of inhomogenous strain, inhomogenous In composition, and surface recombination and their dependencies on the column diameter. Finally, broad-band "white" PL emission was demonstrated from a single InGaN growth on an ensemble of GaN nanocolumns with a broad distribution of column diameters.

REFERENCES

1. E. Calleja et al., Phys. Rev. B **62**, 16826 (2000).
2. M. T. Björk et al., Appl. Phys. Lett. **80**, 1058 (2002).
3. C. Rivera et al., Phys. Rev. B **75**, 045316 (2007).
4. A. Kikuchi, M. Kawai, M. Tada, and K. Kishino, Jpn. J. Appl. Phys. **43**, L1524 (2004).
5. K. A. Bertness et al., J. Cryst. Growth **287**, 522 (2006).
6. C. L. Hsiao et al., J. Vac. Sci. Technol. B **24**, 845 (2006).
7. R. Calarco et al., Nano Lett. **5**, 981 (2005).
8. H.-Y. Chen et al., Appl. Phys. Lett. **89**, 243105 (2006).
9. J. Abell and T.D. Moustakas, Appl. Phys. Lett. **92**, 091901 (2008).
10. K. S. Ramaiah et al., Appl. Phys. Lett. **85**, 401 (2004).
11. C. Wetzel and T. Detchprohm, MRS Internet J. Nitride Semicond. Res. **10**, 2 (2005) and references therein
12. Y. Kawakami et al., Appl. Phys. Lett. **89**, 163124 (2006).

Mater. Res. Soc. Symp. Proc. Vol. 1144 © 2009 Materials Research Society 1144-LL16-06

Near-Infrared Lasers in GaAs/GaAsP Coaxial Core-Shell Nanowires

Bin Hua, Junichi Motohisa, Shinjiroh Hara, and Takashi Fukui

Research Center for Integrated Quantum Electronics (RCIQE) and Graduate School of Information Science and Technology, Hokkaido University, Sapporo 060-8628, Japan

ABSTRACT

Highly uniform GaAs/GaAsP coaxial nanowires were prepared via selective-area metal organic vapor phase epitaxy (SA-MOVPE). Photoluminescence (PL) spectra from a single nanowire indicate that the obtained heterostructures can produce near-infrared (NIR) lasing under pulsed light excitation. The end facets of a single nanowire form natural mirror surface to create an axial cavity, which realizes resonance and give stimulated emission. This study is a considerable advance towards the realization of nanowire-based NIR light sources.

INTRODUCTION

Interest in one-dimensional semiconductor nanowires continues to grow since they have led to great development in the potential of optoelectronics or photonic devices. In recent several years, semiconductor subwavelength nanowires have been demonstrated to show laser emission [1-4]. Representative semiconductor materials fabricating nanowire lasers involve ZnO, GaN and CdS etc. Such nanowire lasers are currently among the smallest known lasing devices, with lengths between one and several tens micrometers and diameter that can be significantly smaller than the emission wavelength in vacuum. In this size range, because of large dielectric contrast between the nanowire and the ambient, a strong lateral optical confinement is created. For a single-crystalline nanowire, the end facets form natural mirror surface that create an axial resonator. That is, one-dimensional semiconductor nanowires not only act as a gain medium but also a waveguide and a Fabry-Perot resonator, which provide coherent feedback.

However, most advances of nanowire lasers are successfully realized via wide-bandgap semiconductor materials, giving an ultraviolet to blue laser emission. To date limited investigation of nanowire lasers in near-infrared (NIR) spectral range is reported [5], while there are vast areas of applications in NIR lasers particularly for optical fiber communications. For instance, lasers operating at 850 nm can be used in-house and in-enterprise local area network with extremely high transmission bit rate. In this case, GaAs-based materials are primarily important while the difficulty to realize nanowire lasers in these material systems is that high density of surface states will lead to strong decay of the emission intensity in GaAs nanowires. Furthermore, currently, most nanowires are grown via vapor-liquid-solid growth, using gold as a catalyst [6]. Although the evidence and consequence of the incorporation of gold are not clear (gold is thought to be deep acceptors in GaAs), elimination of such metal catalysts is advantageous to realize high-purity materials and to avoid the possibilities of influence on their optical and electronic transport properties. We recently develop selective-area metal organic vapor phase epitaxy (SA-MOVPE) as a versatile method to realize well controlled nanowires without any catalyst. The grown nanowires exhibit uniform, flat and smooth morphology as well as good controllability to form heterostructures [7,8]. Here we present GaAs/GaAsP coaxial core-shell nanowires fabricated by SA-MOVPE and observe that single wire shows optically pumped NIR lasing emission.

EXPERIMENTAL

SA-MOVPE growth of GaAs nanowires started with the deposition of 30 nm SiO_2 layer by plasma sputtering on (111)B GaAs substrate. Then SiO_2 was partially removed by electron beam lithography and wet chemical etching. The mask pattern of SiO_2 had the opening holes arranged in a triangular lattice. The diameter of the grown nanowires was directly related to that of the opening holes. The patterned substrates were loaded into a horizontal low-pressure MOVPE system working at 0.1 atm using trimethylgallium (TMG), tertiarybutylphosphine (TBP) and arsine (AsH_3) as source materials. GaAs and GaAsP were grown in succession for core-shell heterostructures. For GaAs, the partial pressures were 1.0×10^{-6} and 2.5×10^{-4} atm for TMG and AsH_3, respectively. The growth temperature and growth time were set to 750 °C and 60 min. In the growth of GaAsP shell, the partial pressures were 1.0×10^{-6}, 2.2×10^{-4} and 2.5×10^{-4} atm for TMG, TBP and AsH_3, respectively. The growth temperature and growth time were 650 °C and 5 min. With these conditions, perfect selective epitaxy of both GaAs and GaAsP was achieved.

For single nanowire optical characterization, as-grown nanowires were mechanically cut down and dispersed onto 2-μm-thick-SiO_2-covered Si substrates. Photoluminescence (PL) measurements were performed using a regeneratively amplified Ti: sapphire pulsed laser (140 fs pulse duration, 78 MHz repetition rate) with a wavelength of 753 nm. The excitation beam was focused onto ~ 2 μm in diameter with a ×50 microscope objective (NA = 0.42) on the sample placed in a cryostat. The emission through the same microscope objective was collected into a liquid-nitrogen-cooled charge coupled device (CCD) for spectral analysis or a CCD camera for imaging.

DISCUSSION

The overall morphology of as-grown nanowires was examined by scanning electron microscopy (SEM). Typical SEM images of a growth of GaAs/GaAsP core-shell nanowires are shown in Figure 1a. A uniform array of vertical standing nanowires is prepared with smooth top surface and sidewall. Top view images of a nanowire show a clear hexagonal cross section. The wires grow perpendicularly on GaAs (111)B wafers, thus the growth direction coincides with the <111> direction. Since this SA-MOVPE method fully utilizes the nature of epitaxial growth and avoids the use of gold as the seed for nanowires, the nanowires are synthesized with superior crystalline quality with atomic precision. In this study, nanowire diameters, d, were controlled to range between 200 and 500 nm, and lengths ranged between 2 and 6 μm. The nanowire size depends on the prepared opening hole on substrates and growth conditions. The relatively larger nanowire diameters in this experiment are necessary to enhance reflectivity at both end surface and bring sufficient optical confinement for lasing, especially at NIR wavelengths. Figure 1b and c show a SEM comparison of bare GaAs and GaAs/GaAsP core-shell nanowires. The length of the core-shell wire is almost equal to that of the GaAs one, but the diameter of the former is approximately 100 nm larger than that of the latter, suggesting a 50 nm thick GaAsP shell around GaAs core wire.

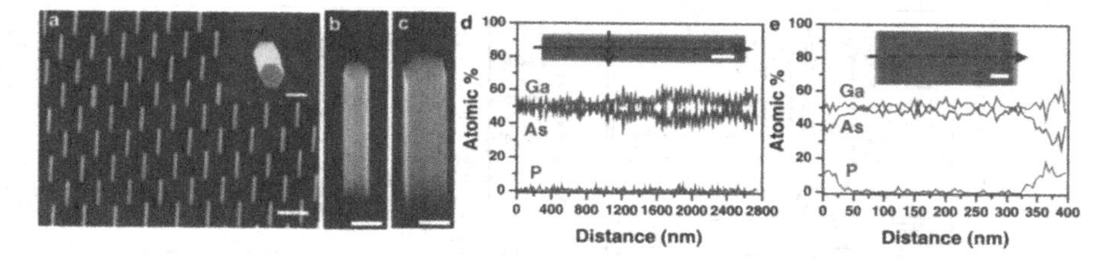

Figure 1. (a) Scanning electron microscopy (SEM) images of well-aligned GaAs/GaAsP core-shell nanowires arrays grown on GaAs (111)B substates. Scale bar, 2 μm. Inset shows a top view image of a nanowire with hexagonal cross section. Scale bar, 200 nm. (b) and (c) SEM images of as-grown GaAs and GaAs/GaAsP nanowires for comparison. Scale bars, 200 nm. (d) A bright-field TEM image and axial EDXS linescan of a single nanowire showing variations in Ga, As and P compositions. Scale bar, 300 nm. (e) The TEM image and radial EDXS scan along the dotted line arrow in d. Scale bar, 50 nm.

To examine the wire structure and composition, transmission electron microscopy (TEM) and energy dispersive x-ray spectroscopy (EDXS) were then applied. A bright-field TEM image and EDXS linescans obtained from a single nanowire axial plane exposed by focused ion beam etching are shown in Figure 1d and e. The length and diameter of this nanowire are 2.8 μm and 390 nm, respectively. In Figure 1d, results are shown for an EDXS linescan along the axis of the nanowire (along the solid line arrow). Another linescan performed across the diameter of the nanowire is shown in Figure 1e. The scan position is labeled with a dotted line arrow in Figure 1d. The axial EDXS scan indicates that the nanowire core is entirely composed of GaAs along the center axis and P content on its top surface can be negligible. However the radial linescan clearly shows presence of P element around the GaAs core. These results prove that the nanowire structure should be a GaAs core surrounded by a GaAsP shell. The P concentration in GaAsP shell is around 10 % in atomic, i.e. $GaAs_{0.8}P_{0.2}$, and the thickness of outer GaAsP shell is estimated to be 50 nm, which is consistent with that SEM observation before and after the growth of GaAsP.

In our previous study, Fabry-Perot microcavity modes have been observed in single GaAs nanowires [9]. PL spectra results suggest that a Fabry-Perot microcavity is formed along the length of the nanowire and the (111) facets of both ends act as reflecting mirrors. But due to nonradiative recombination of photoexcited carriers at the air-exposed GaAs sidewall surface where high density of surface states exists, the emission intensity of bare GaAs wires is too weak to obtain lasing. For further optimization of this cavity quality, core-shell nanowires are expected instead of bare GaAs wires. The function of the shell layer is passivation of the surface states of the GaAs core nanowires, resulting in the high-quality and optically active nanowires. In this study, PL spectra of GaAs/GaAsP core-shell nanowires display apparently strong emission intensity, which is stronger than that of bare GaAs nanowires by a factor of over two orders of magnitude.Here the shell in 50 nm thickness should be sufficient to passivate surface states since it was reported that a shell in thickness around 10 nm suppressed the reduction of PL efficiency [10,11].

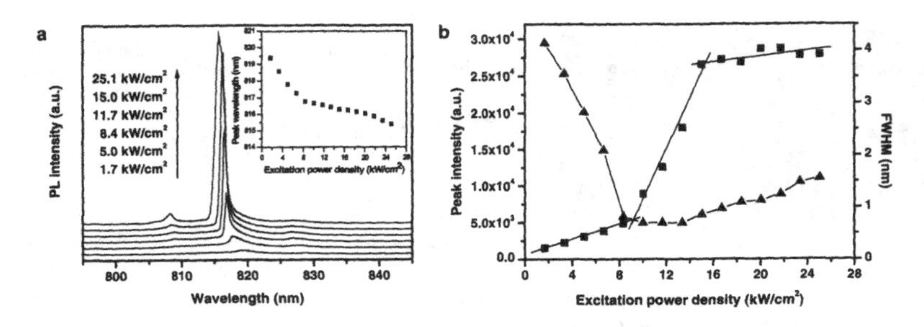

Figure 2. (a) PL spectra from a single GaAs/GaAsP core-shell nanowire as a function of increasing excitation power density at 4.2 K. Inset shows excitation power density dependent lasing peak wavelength. (b) Plots of emission peak intensity (■) at the center of 816 nm and full width at half maximum (FWHM) (▲) of PL spectra versus excitation power density.

Then an isolated single GaAs/GaAsP core-shell nanowire was excited with a pulsed laser and PL emission were subsequently collected at 4.2 K, displayed in Figure 2a. This wire is 330 nm in diameter and 5.5 μm in length. At low excitation power densities PL spectra display a broad emission band at about 820 nm. However, above a certain threshold (8.4 kW/cm^2 in this nanowire), a sharp and narrow peak centered at ~ 816 nm appears and the peak intensity increases rapidly with excitation energy. This emission peak is exactly consistent with GaAs band gap energy (~ 1.52 eV at 4.2 K), indicating the onset of stimulated emission from the GaAs core of nanowire. From Figure 2b, stimulated emission or lasing is evidenced by the appearance of the simultaneous line narrowing and superlinear increase of intensity at pump densities above the threshold. But with further increasing pump power density above 15.0 kW/cm^2, the emission intensity saturates, indicating that the gain is pinned. We speculate the quick saturation of laser emission is in the present experimental configuration. We use tightly focus of incident laser on nanowire, and the diameter of laser beam is around 2 μm, which is much less than the wire length. These lead to a smaller pumping area in nanowire, thus the gain is obtained only in the vicinity of excitation region. As a result, gain saturates quickly. Excitation-induced local heating may also a reason for quick gain saturation. Moreover, for conventional semiconductor lasers, the line width varies inversely with pump power, due to an increasing degree of population inversion. However, microlasers pumped with short intense pulses can exhibit additional nonequilibrium effects that act to broaden the lasing peak [12]. Such effect is also seen in Figure 2b. The laser peak is broadened from 0.7 to 1.5 nm over the power range of 13.4 to 25.1 kW/cm^2. It is likely that there is increased spontaneous emission noise at higher pump intensities due to amplified spontaneous emission that causes phase fluctuations that couple to the real and imaginary parts of the material susceptibility. These optical density fluctuations would couple to the cavity modes and act to broaden the lines.

In usual, nanowire lasing becomes redshifted with the increasing excitation power, due to heating effect or the band gap renormalization induced by the electron-hole plasma state [13]. Contrarily, a blueshift of lasing peak in this wire is observed by an increase of pump power in Figure 2a inset. The reason of this blueshift is unclear but we propose that one possibility may be due to bandbending [14]. Upon excitation, photoexcited charge carriers are generated and reach a dynamic equilibrium, where electrons get confined at the nanowire surface but holes accumulate at the nanowire center. Thus an electron near the surface recombines with a hole in the center, leading to low transition energy, while at the high excitation intensity, the bandbending effect weakens due to the formation of more charge carrier, bringing high emission energy and transition probability.

The lasing characteristics of the GaAs core indicate that these heterostructure nanowires are very efficient gain media, which can be attributed to the ideal photon confinement structures. Since the regular dimensions of the wires make them good waveguides, the emission light can be accumulated and amplified through reflecting from both ends of the wires and finally produce stimulated emission or lasing under enough high excitation. The lasing in the present core-shell structures is mainly attributable to the reduction of non-radiative surface recombination as discussed above. As a comparison, it must be emphasized that no lasing has been observed thus far for pure GaAs nanowires.

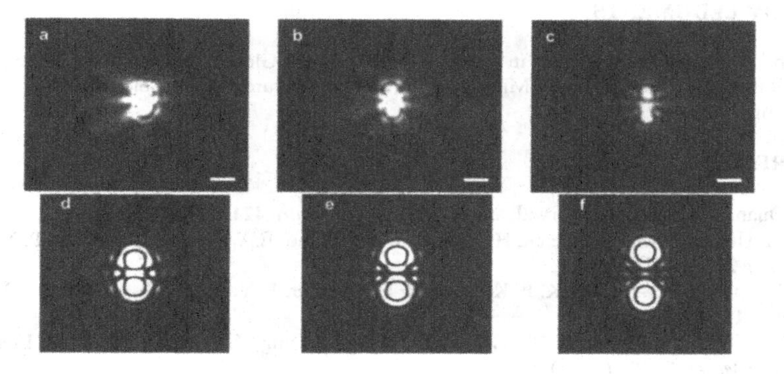

Figure 3. (a-c) Far-field optical images of lasing emission from different nanowires in a length of 3.4, 4.5 and 5.5 μm, respectively. Scale bars, 5 μm. (d-f) Simulated interference patterns for corresponding nanowires in a-c, assuming a spherical non-directional emission from the wire end facets.

Figure 3a-c shows lasing emission images from several core-shell nanowires in different lengths. All wires exhibit very bright luminescence spots observed at both ends. This is a typical feature of an optical waveguide and it suggests that the wires are able to absorb the excitation light and propagate the PL emission towards the ends. Almost all of the core-shell nanowires exhibit this kind of waveguide behavior. Since the waveguided light is generated from PL within the nanowire, these wires can be classified as active waveguides, as compared to those passive ones where light must be coupled-in from external sources. Interestingly, intensity modulations around the wires are observed, suggesting that the diffraction and interference occur, which originate from the spherical emission from both wire end facets with a fixed phase difference. These interference patterns are expected to depend on the wire length. For more quantitative understanding, we simulated diffraction and interference patterns following an approach described in Ref. 15 (assuming spherical non-directional emissions in nanowire end facets) and are presented in Figure 3d-f. Here, NA of the microscope objective and the difference of the phase at two edges are taken into account. One can see that the interference patterns around the nanowires are nicely reproduced for all three nanowires.

CONCLUSIONS

In conclusion, we have presented NIR lasing in GaAs-core one-dimensional semiconductor nanostructures with a GaAsP barrier shell. It is believed that the lower threshold laser could be addressed by further optimizing the cavity quality of core-shell nanowires. One possible method is to prepare Bragg gratings at the nanowire ends through axial composition modulation. We expect GaAs-based nanowire lasers to affect the applications for telecommunications, data storage, and future integrated-photonic platsforms.

ACKNOWLEDGMENTS

This work is financially supported in part by a grant from the Global Center of Excellence (GCOE) Program (No. C01) from Ministry of Education, Culture, Sports, Science and Technology, Japan.

REFERENCES

1. X. Duan, Y. Huang, R. Agarwal, and C. M. Lieber, *Nature* **421**, 241 (2003).
2. M. H. Huang, S. Mao, H. Feick, H. Yan, Y. Wu, H. Kind, E. Weber, R. Russo, and P. Yang, *Science* **292**, 1897 (2001).
3. J. C. Johnson, H. –J. Choi, K. P. Knutsen, R. D. Schaller, P. Yang, and R. J. Saykally, *Nature Mater.* **1**, 106 (2002).
4. F. Qian, Y. Li, S. Gradecak, H. –G. Park, Y. Dong, Y. Ding, Z. L. Wang, and C. M. Lieber, *Nature Mater.* **7**, 701 (2008).
5. A. H. Chin, S. Vaddiraju, A. V. Maslov, C. Z. Ning, M. K. Sunkara, and M. Meyyappan, *Appl. Phys. Lett.* **88**, 163115 (2006).
6. R. S. Wagner, and W. C. Ellis, *Appl. Phys. Lett.* **4**, 89 (1964).
7. L. Yang, J. Motohisa, J. Takeda, K. Tomioka, and T. Fukui, *Appl. Phys. Lett.* **89**, 203110 (2006).
8. P. Mohan, J. Motohisa, and T. Fukui, *Appl. Phys. Lett.* **88**, 133105 (2006).
9. B. Hua, J. Motohisa, Y. Ding, S. Hara, and T. Fukui, *Appl. Phys. Lett.* **91**, 131112 (2007).
10. J. M. Moison, K. Elcess, F. Houzay, J. Y. Marzin, J. M. Gerard, F. Barthe, and M. Bensoussan, *Phys. Rev. B* **41**, 12945 (1990).
11. S. Kodama, S. Koyanagi, T. Hashizume, and H. Hasegawa, *J. Vac. Sci. Technol.* **13**, 1794 (1995).
12. J. C. Johnson, H. Yan, P. Yang, and R. J. Saykally, *J. Phys. Chem. B* **107**, 8816 (2003).
13. P. J. Pauzauskie, D. J. Sirbuly, and P. Yang, *Phys. Rev. Lett.* **96**, 143903 (2006).
14. M. H. M. van Weert, O. Wunnicke, A. L. Roest, T. J. Eijkemans, A. Y. Silov, J. E. M. Haverkort, G. W. 't Hooft, and E. P. A. M. Bakkers, *Appl. Phys. Lett.* **88**, 043109 (2006).
15. L. K. van Vugt, S. Ruhle, and D. Vanmaekelbergh, *Nano Lett.* **6**, 2707 (2006).

Mater. Res. Soc. Symp. Proc. Vol. 1144 © 2009 Materials Research Society 1144-LL17-03

Fabrication of ZnO Bridging Nanowire Device by a Single-Step Chemical Vapor Deposition Method

Yanbo Li, Ippei Nagatomo, Ryohei Uchino, Ichiro Yamada, and Jean-Jacques Delaunay

Department of Engineering Synthesis, The University of Tokyo, 7-3-1 Hongo, Bunkyo-ku, Tokyo 113-8656, Japan

ABSTRACT

ZnO nanowires are directly integrated into a working device by a single-step chemical vapor deposition (CVD) method. Gold catalyst is patterned on a quartz glass substrate using a comb-shaped shadow mask and then ZnO is grown on the patterned substrate by CVD. Thick ZnO layers formed on the gold-patterned areas serve as native electrodes. Ultra-long (~100 μm) ZnO nanowires grown across the gap between the ZnO electrodes and the nanowires serve as the sensing elements of the device. The device exhibits high sensitivity and fast response to UV illumination in air. Our method can be used to fabricate other metal oxide semiconductor bridging nanowire devices, which have promising applications in photodetection and gas sensing.

INTRODUCTION

Zinc oxide is a direct, wide bandgap (E_g ~ 3.4 eV) semiconductor material with a wide range of properties from UV detection to gas sensing [1-5]. The sensing properties of ZnO and other metal oxides are governed by the charge-exchange reactions with adsorbants on the surface [6]. Due to their inherently high surface-to-volume ratio, ZnO nanowires offer the promise of high sensitivity. Although the nanowires can be grown by simple thermal evaporation or chemical vapor deposition (CVD) methods with relatively low cost [7-9], the integration of the nanowires into working devices still remains a challenge. The integration of nanowires is usually done by "bottom-up" and then "top-down" processes. The nanowires are grown by "bottom-up" processes such as thermal evaporation or CVD. Then, they are collected from substrates and dispersed in a solution. After that, the nanowires are randomly deposited on an insulating substrate. The positions of the nanowires are noted under scanning electron microscope (SEM) observation with the assistance of pre-patterned markers on the substrates. Thereafter, metallic contacts to the nanowires are fabricated by "top-down" techniques such as photolithography [10], electron beam lithography [11], or focused ion beam lithography [12]. Because of its complexity, low efficiency and high cost, this method is limited to fundamental studies of properties of nanowires. In recent years, a new process has been developed to integrate nanowires into working devices. It is based on the concept of bridging nanowires between electrodes by self-organized growth of nanowires [13-19]. On contrary to the first method, the bridging method employs a "top-down" process to make the electrodes, and then uses a "bottom-up" process to grow the nanowires. A single crystal substrate is usually used to make a trench and form electrodes on both sides of the trench by a "top-down" process. Then, nanowires are grown across the trench from one electrode to the other, forming bridges directly in the "bottom-up" growth process. The integration process is more efficient in comparison to the first method. However, expensive single crystal substrates are used to make the trench so as to achieve epitaxial nanowire growth and thus form ohmic contacts between the trench walls and the nanowires. Besides, "top-down" techniques such as photolithography, lift-off, and reactive-ion-etching are still needed to fabricate the trenches. Here, we report the fabrication of a ZnO bridging nanowire device by a single-step CVD method without resorting to any "top-down" process. In our method, the electrodes and the sensing elements are formed simultaneously during the CVD process. Despite of the simplicity of the fabrication process, our device is extremely sensitive to UV illumination.

EXPERIMENT

The schematic illustration of the experimental setup for growing ZnO bridging nanowires is shown in Figure 1. The device was fabricated by a CVD process in a horizontal vacuum tube furnace. A thin (~2 nm) gold discontinuous layer was sputtered on a quartz glass substrate through a comb-shaped shadow mask. The number of comb fingers was 25, their length was 5 mm, and the width of the fingers and the gap between them were both 100 μm. A powder mixture of 0.6 g ZnO (Koch Chemicals, 99.999 % purity) and 0.3 g graphite (NewMet Koch, 99.999 % purity) was charged in an alumina boat located at the center of the furnace. The substrate was placed 4-6 cm away from the source and downstream of the carrier gases. Argon and oxygen with a ratio of 10:1 were used as the carrier gases at a working pressure of ~70 mbar. The temperature of the source was kept at 1050 °C for 30 min, while the temperature of the substrate was around 1000 °C. Then, the furnace was cooled down at a rate of 10 °C min^{-1} to room temperature. SEM images of the sample were taken with a Hitachi S3000N. For electrical measurements, copper wires were fixed by indium granules on both pads of the ZnO combs. The time-dependent photoresponse was measured using a source meter (Keithley, 2400) connected with a GPIB controller to a computer. A mercury arc lamp was used as the UV light source, and an excitation filter centered at 350 nm (±30 nm) was inserted in the beam. The UV irradiance was varied using neutral density filters and measured by an optical power meter (Ophir PD300-UV). A mechanical shutter was used to turn on and off the UV illumination on the sample.

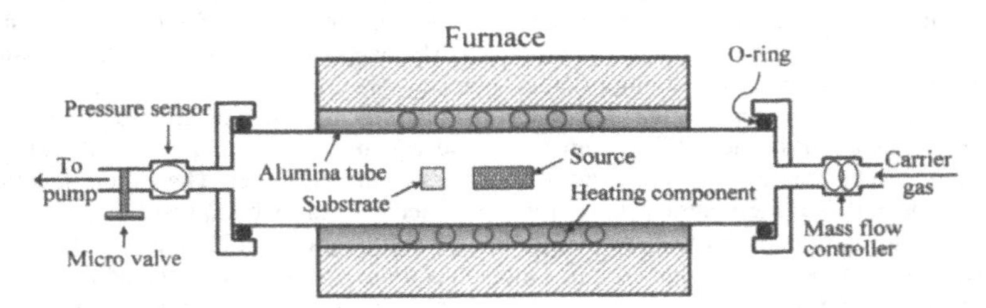

Figure 1. Schematic illustration of the experimental setup (top view) for the growth of ZnO bridging nanowires.

DISCUSSION

The idea of fabricating ZnO bridging nanowire devices was inspired by the observation of different morphologies of ZnO products on different areas of the substrate. In our preliminary experiments, it was found that under the CVD conditions described in the experiment section, a thick ZnO layer was formed on top of a quartz glass substrate sputtered with a layer of gold (2 nm in thickness). Under SEM observation, it was found that this thick layer consisted of tangling ZnO nanowires and nanosheets, as shown in Figure 2a. The thickness of the layer is of the order of tens of microns and the resistance of the layer is very low according to our measurements. At the edges of the substrate, ZnO nanowires grow very long and are quite uniform because there is enough space for the nanowires to grow laterally, as exhibited in Figure 2b. The diameter of the nanowires is about 100 nm and their length around 50-100 μm. These results are very interesting because if two thick ZnO layers are close enough to each other, the long and oriented nanowires at their edges can bridge between them. Thus direct integration of nanowires into a working device is made possible, because the thick layers may serve as electrodes and the bridging nanowires can work as sensing elements.

Based on this idea, we have successfully fabricated ZnO bridging nanowire device by a single-step CVD process using quartz glass substrate sputtered with Au discontinuous thin film with defined shape. As shown in Figure 3a, ZnO thick layer was formed selectively on the gold sputtered areas, that is, the pads and the fingers of the comb-shaped structures. The morphology

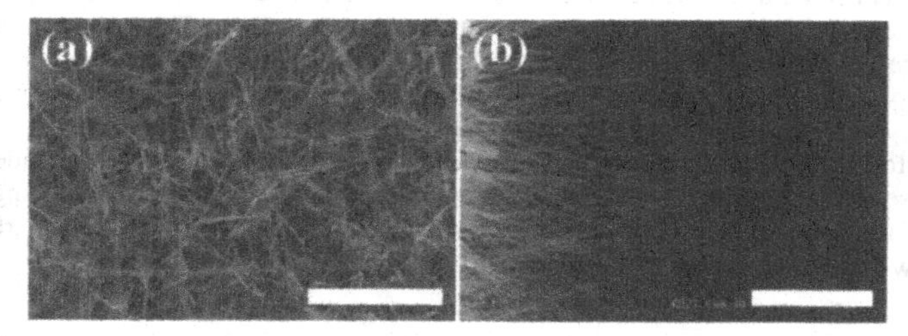

Figure 2. SEM images of ZnO nanostructures grown (a) on top of the quartz glass substrate and (b) at the edge of the substrate. The scale bars are both 20 μm.

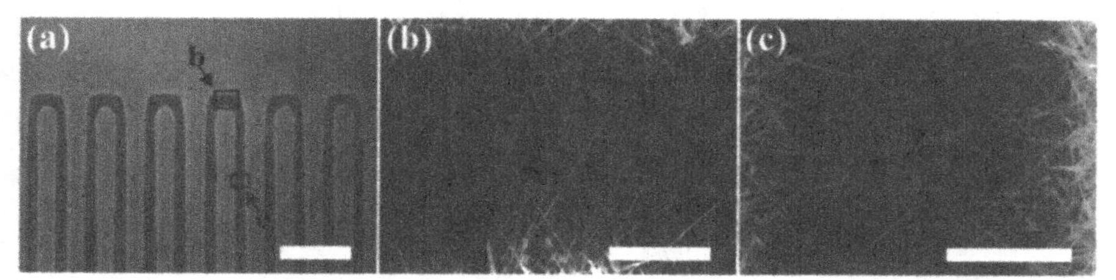

Figure 3. SEM images of the fabricated ZnO bridging nanowire device. The positions of images (b) and (c) are outlined in black in (a). The scale bars are (a) 500 μm, (b) 30 μm, and (c) 30 μm.

of the layer is the same as that shown in Figure 2a. Figure 3b and 3c show the details of the gap between the comb fingers. It can be seen that the gap between the fingers are bridged by some long nanowires. Because the resistance of the bridging nanowires is much larger than that of the thick layers on the pads and fingers, the sensing performance should mainly depend on the bridging nanowires. Due to the large surface-to-volume ratio of the bridging nanowires, the device is promising for high sensitivity gas detection and photon detection. The two interlacing combs made of thick ZnO layers serve as electrodes to collect electrical signals. The electrodes and the sensing elements are made simultaneously in the CVD process using our method, which makes the fabrication process very simple and highly efficient. Besides, the contamination to the nanowire surface is also minimized because no post treatment is needed to fabricate electrodes.

Selective growth of ZnO thick layer on the substrate is the key factor for achieving this bridging nanowire device. Thus, the selection of catalyst and substrate is very important for our method. Au atoms have a strong binding affinity to ZnO [20], so that ZnO particles in the gas flow can be caught easily by Au catalysts on the substrate. Moreover, the quartz glass has a very low surface energy of ~1.8 J m^{-2} [21], so that ZnO nucleation events do not take place without the presence of Au on the quartz glass substrate. Although the exact role of Au in the growth of the bridging nanowires is still unclear, three possible growth modes have been reported that may take place at the same time. First, the nanowire growth follows a vapor-liquid-solid (VLS) growth, where the Au nanoparticles act as nuclei for the nanowire and subsequently as growth sites. This is evidenced by the observation of an Au droplet at the tip of some of the ZnO nanowires. Second, the nanowires grow from the ZnO nuclei formed on the Au nanoparticles [22]. In this case, the Au nanoparticles are at the root of the nanowires. Third, the nanowires grow from the ZnO layers formed on top of Au layer. This type of nanowire growth follows a catalyst free process. Catalyst free growths of ZnO nanostructures usually involved ZnO seeds [23], so this type of growth is also possible for our sample.

The performance of the ZnO bridging nanowire device was tested by time-dependent photoresponse measurements. Figure 4 shows the photoresponse measured by switching on and off the UV illumination repeatedly over six cycles. Under a bias voltage of 5 V, the dark current level is below 10^{-9} A. Upon UV illumination, the current rises immediately to ~2×10^{-5} A. The

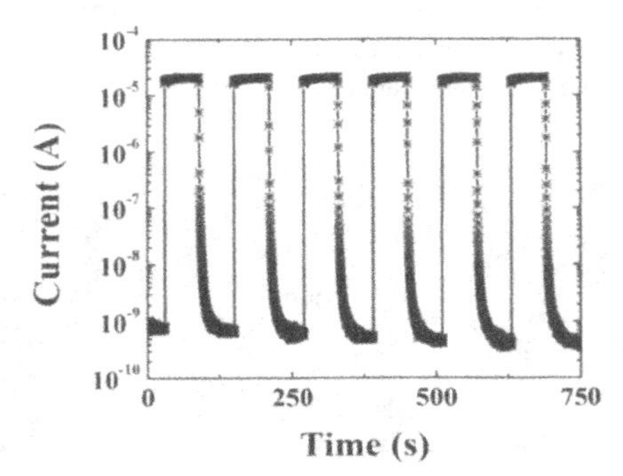

Figure 4. Photoresponse of the ZnO bridging nanowire device upon UV illumination switching on (for 60 s) and off (for 60 s) over six cycles. The bias voltage was 5 V and the irradiance was 0.8 mW cm^{-2}.

current increases by more than four orders of magnitude under an irradiance of 0.8 mW cm^{-2}. After switching off the UV illumination, the current decreases quickly to the dark level. As can be seen from the curve, the current fully recovers within 60 s, which is much faster than most reported results. By fitting the current decay data with an exponential decay formula ($y = y_0 + Ae^{-x/t}$), decay time for the six decays was obtained, giving an average decay time of only 0.3 s. The photoresponse also shows high reproducibility as evidenced by the nearly identical rises and decays over the six cycles. The above results demonstrate that the ZnO bridging nanowire device is highly sensitive to UV illumination. Our fabrication method provides a simple and cost-effective way to make high-performance UV detectors. The detailed UV sensing properties of the ZnO bridging nanowires have been reported elsewhere [24, 25].

In order to prove that the high sensitivity of the device is truly attributed to the bridging nanowires, we have measured the photoresponse of the thick ZnO layers made of tangling ZnO nanowires and nanosheets. As shown in Figure 5, the current increases by four orders of magnitude upon UV illumination for photoresponse measured between the two interlaced combs. However, the current increases by only four times upon UV illumination when the photoresponse is measured between the points B and C of the same comb pad (see Figure 5). Furthermore, the current recovery of the bridging nanowires (points A and B) is faster than that of the thick layer (points B and C). Therefore, the high sensitivity and fast response of the device is ascribed to the nanowires bridging the fingers of the interlaced combs.

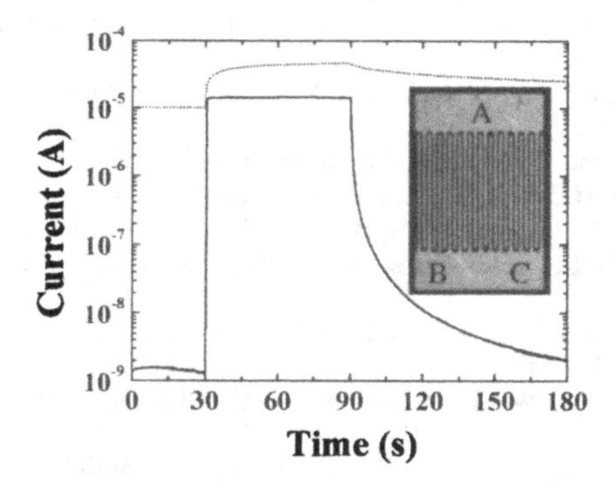

Figure 5. Photoresponse measured under the same UV illumination (0.77 mW cm^{-2}) and bias voltage (5 V) in air between the points A and B (solid line), and between the points B and C (dotted line). The location of the points A, B, and C is indicated in the inset showing an optical microscope image of the fabricated device.

CONCLUSIONS

In conclusion, we have fabricated a ZnO bridging nanowire device by a single-step CVD method. The electrodes and the sensing elements were fabricated simultaneously in the CVD process, which is due to selective growth of thick ZnO layers on Au catalyst covered areas and lateral growth of ultra-long ZnO nanowires at the edges of the thick layers. The device exhibited high sensitivity, fast response, as well as high reproducibility to UV illumination. The high performance of the device was proved to be attributed to the bridging nanowires. The device is also expected to have high sensitivity to gas sensing. Our one-step growth method provides a simple, efficient, and cost-effective way to fabricate UV or gas sensors for practical applications.

ACKNOWLEDGMENTS

This work was partly supported through the Global COE Program, "Global Center of Excellence for Mechanical Systems Innovation," by the Ministry of Education, Culture, Sports, Science and Technology.

REFERENCES

1. H. Kind, H. Yan, B. Messer, M. Law, and P. Yang, *Adv. Mater.* **14**, 158 (2002).
2. M. H. Huang, S. Mao, H. Feick, H. Yan, Y. Wu, H. Kind, E. Weber, R. Russo, and P. Yang, *Science* **292**, 1897 (2001).
3. S. Ju, A. Facchetti, Y. Xuan, J. Liu, F. Ishikawa, P. Ye, C. Zhou, T. J. Marks, and D. B. Janes, *Nature Nanotech.* **2**, 378 (2007).
4. R. L. Hoffman, B. J. Norris, and J. F. Wager, *Appl. Phys. Lett.* **82**, 733 (2003).
5. J.-J. Delaunay, I. Nagatomo, R. Uchino, Y. B. Li, M. Shuzo, and I. Yamada, *Adv. Mater. Res.* **47**, 634 (2008).
6. Y. Takahashi, M. Kanamori, A. Kondoh, H. Minoura, and Y. Ohya, *Jpn. J. Appl. Phys.* **33**, 6611 (1994).
7. S. C. Lyu, Y. Zhang, and C. J. Lee, *Chem. Mater.* **15**, 3294 (2003).
8. M. H. Huang, Y. Wu, H. Feick, N. Tran, E. Weber, and P. Yang, *Adv. Mater.* **13**, 113 (2001).
9. Y. Li, M. Zheng, L. Ma, M. Zhong, and W. Shen, *Inorg. Chem.* **47**, 3140 (2008).
10. Z. Fan, P. Chang, J. G. Lu, E. C. Walter, R. M. Penner, C. Lin, and H. P. Lee, *Appl. Phys. Lett.* **85**, 6128 (2004).
11. C. Soci, A. Zhang, B. Xiang, S. A. Dayeh, D. P. R. Aplin, J. Park, X. Y. Bao, Y. H. Lo, and D. Wang, *Nano Lett.* **7**, 1003 (2007).
12. X. Zhang, Y. Zhang, Z. L. Wang, W. Mai, Y. Gu, W. Chu, and Z. Wu, *Appl. Phys. Lett.* **92**, 162102 (2008).
13. K. Haraguchi, K. Hiruma, T. Katsuyama, K. Tominaga, M. Shirai, and T. Shimada, *Appl. Phys. Lett.* **69**, 386 (1996).
14. M. S. Islam, S. Sharma, T. I. Kamins, and R. S. Williams, *Nanotechnology* **15**, L5 (2004).
15. J. S. Lee, M. S. Islam, and S. Kim, *Sens. Actuators B* **126**, 73 (2007).
16. J. F. Conley, L. Stecher, and Y. Ono, *Appl. Phys. Lett.* **87**, 223114 (2005).
17. R. S. Chen, S. W. Wang, Z. H. Lan, J. T. H. Tsai, C. T. Wu, L. C. Chen, K. H. Chen, Y. S. Huang, and C. C. Chen, *Small* **4**, 925 (2008).
18. J. S. Lee, M. S. Islam, and S. Kim, *Nano Lett.* **6**, 1487 (2006).
19. R. He, D. Gao, R. Fan, A. I. Hochbaum, C. Carraro, R. Maboudian, and P. Yang, *Adv. Mater.* **17**, 2098 (2005).
20. K. Albert-Polacek and E. F. Wassermann, *Thin Solid Films* **37**, 65 (1976).
21. Y. K. Shchipalov, *Glass Ceram.* **57**, 374 (2000).

22. D. S. Kim, R. Scholz, U. Gösele, and M. Zacharias, *Small*, **4**, 1615 (2008).
23. W. I. Park, D. H. Kim, S. W. Jung, and G. C. Yi, *Appl. Phys. Lett.* **80**, 4232 (2002).
24. Y. Li, F. Della Valle, M. Simonnet, I. Yamada, and J.-J. Delaunay, *Nanotechnology* **20**, 045501 (2009).
25. Y. Li, F. Della Valle, M. Simonnet, I. Yamada, and J.-J. Delaunay, *Appl. Phys. Lett.* **94**, 023110 (2009).

Mater. Res. Soc. Symp. Proc. Vol. 1144 © 2009 Materials Research Society 1144-LL18-06

Modelling of the Oxidation of Suspended Silicon Nanowires

P. F. Fazzini[1-2], C. Bonafos[1], A. Hubert[3], J.-P. Colonna[3], T. Ernst[3], M. Respaud[4], F. Gloux[1]

[1] *CEMES-CNRS* - Université de Toulouse,29 rue J. Marvig, 31055, Toulouse, France

[2] LAAS-CNRS - Université de Toulouse, 7 avenue du Colonel Roche, 31077, Toulouse, France

[3] CEA-LETI, Minatec, 17 rue des martyrs, 38054 Grenoble Cedex 9, France

[4] LPCNO, INSA, Département de Physique, 135 avenue de Rangueil, 31077, Toulouse France

ABSTRACT

The oxidation of suspended Si nanowires is studied under wet and dry conditions. The nanowire characteristics are extracted from Electron Microscopy images. In parallel, the Deal and Grove model is extended to cylindrical geometry. The used model also assumes that stress effects reduce the oxidation rate and predicts the retardation of oxide growth on curved surface, leading to a self-limited process. The model predictions show a good agreement with experiments.

INTRODUCTION

Ultra-thin and narrow channels (below 10 nm) are ideal for OFF state leakage current (I_{OFF}) control as future sub-22 nm Metal Oxide-Semiconductor (MOS) transistors [1]. Suspended nanowires (NWs) and multi-channels are currently being developed in order to increase the current density per surface unit [2]. The three dimensional (3D) multi-channel Gate-All-Around (GAA) process flow detailed in ref. [3] was used in this study to obtain 1-level suspended Si nanowires. First, 45 nm wide suspended silicon nanowires with near square cross-sections were obtained thanks to e-beam lithography and specific dry etching techniques [4, 5]. Due to optical and electronic lithography limitations, a specific oxidation process has been developed in order to reduce Si NW dimensions to values below 10 nm and to round off the nanowires for optimal I_{OFF} reduction. The use of thermal oxidation to further reduce the diameter of NWs has already been reported [1, 5, 6] but there are only a few data concerning the oxidation kinetic process itself.

The aim of this work is to study and model the thermal oxidation of Si NWs. The methodology and physical hypothesis are similar to what we reported for spherical nanocrystals (NCs) [7]. NW oxidation is examined as a function of the duration of thermal treatments performed at 950°C and 1100°C under respectively wet and dry conditions. The NW characteristics (size of the core and of its oxide envelope) have been extracted from Scanning Electron Microscope (SEM) and Transmission Electron Microscope (TEM) images. In addition, we have extended the

1D Deal and Grove model [8] to cylindrical geometry in order to properly reproduce the Si NW evolution under oxidation. Our model assumes that stress effects associated with non-uniform deformation of the oxide by viscous flow reduce the oxidation rate as observed by Kao et al [9]. This retardation effect increases when the NWs size decreases, leading to a self-limiting oxide growth. The model predictions show a good agreement with the experimental results.

EXPERIMENTAL DETAILS

Both wet and dry oxidations (at 950°C and 1100°C, respectively) have been investigated, for duration up to 20 minutes. The 950°C wet oxidation has been implemented in a vertical furnace using H_2+O_2. The 1100°C dry oxidation has been carried out in a Thermal Rapid Processing (RTP) furnace under O_2 atmosphere. The oxidation duration varies between 300 and 1020 s for the wet process and ranges between 100 and 1200 s for the dry one. The characteristics of the oxidized NWs were measured by SEM. In order to obtain both (i) the dimensions and shape of the NW cross-section and (ii) the thickness of the oxide formed around the NWs, a 25 nm thick Low Pressure Chemical Vapor Deposition (LPCVD) Si_3N_4 encapsulation was carried out. To differentiate the oxide layer from the Si and Si_3N_4 layers using SEM, the samples were also dipped in a HF solution. Fig. 1 shows a cross-sectional TEM image of a Si NW before and after dry oxidation at 1100°C for 600 s. The geometry of the NWs can be described by a pillar with a trapezoidal to rectangular cross-section before oxidation and for short oxidation duration (<250 s).

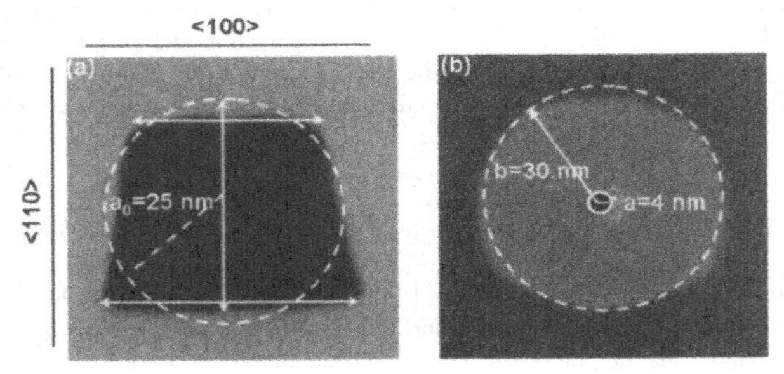

Fig. 1: Cross-sectional TEM images of Si NWs **(a)** for a reference sample (non oxidized)with initial average radius (a_0) of 25 nm and **(b)** for the same sample oxidized under dry conditions at 1100°C for 600 s.

On the contrary, the shape of the NWs can be fitted by a cylinder i. e., a pillar with circular cross-section after longer oxidation duration [10]. When the NWs present a rectangular cross-section, the sidewalls of the rectangles are oriented following <110> and <100>.

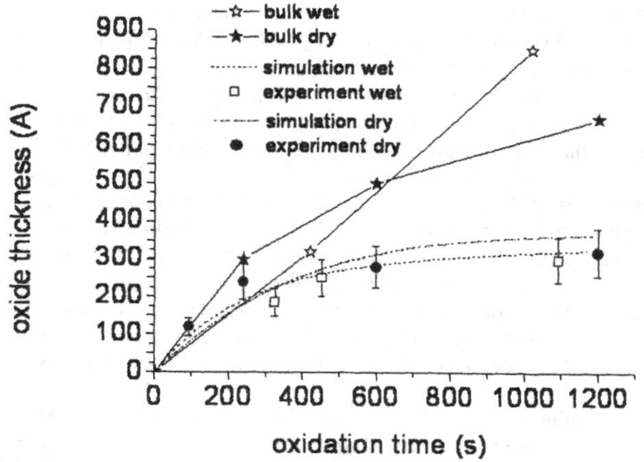

Fig. 2: Comparison between the experimental oxide envelope thickness measured on SEM images (wet oxidation at 950°C: empty squares, dry oxidation at 1100°C: open circles) and the simulation predictions (wet oxidation: dash line, dry oxidation: dash-dot line). Before oxidation, the NW has mean lateral dimension of 50 nm and is described by a cylinder with a circular cross-section of radius a_0=25 nm.

The evolution of the oxide envelope thickness upon oxidation, as measured on the cross-sectional images, is shown on Fig. 2, for NWs with initial mean radius (a_0) of 25 nm (full circle symbols for dry oxidation at 1100°C and open squares for wet oxidation at 950°C). We have considered 20 % for the error bars on the size measurement due to microscope resolution. This evolution is retarded compared to those on (100) planar Si bulk (empty and full stars). Both are very similar below 250 s but the oxidation of the NWs is strongly reduced compared to planar bulk Si oxidation for larger oxidation times. In addition, we observe a saturation in the evolution of the oxide envelope thickness, which reaches an asymptote after 10 minutes of annealing in both cases. As a consequence, the oxidation rate is slower when increasing the annealing duration, i. e., when the size of the NW decreases. In other words, the smaller the NW, the smaller the oxidation velocity is. This self-limited oxidation process has been also observed for 2D micrometric and nanometric structures [9-14] and for spherical (3D) NCs [7]. Both dry oxidation at 1100°C and wet oxidation at 950°C give rise to similar oxide envelopes, typically larger than 25 nm after 20 minutes of oxidation.

DISCUSSION

The most comprehensive characterization study of the stress effects during oxidation of non-planar Si structures was reported by Kao et al [9]. These authors have extended the one dimensional stress-free model of Deal and Grove [8] to the oxidation of non-planar concave and convex cylindrical (2-D) micrometric silicon structures. They considered the oxide as an incompressible viscous fluid being driven by a velocity field at the silicon/oxide interface. This assumption is in fact valid above the viscous flow point (950°C) while at low temperatures (below 800°C) SiO_2 behaves as an elastic solid. To apply the fore-mentioned cylindrical model to the geometry of the system under study (a wire with a trapezoidal section whose surfaces are oriented along the <100> and <110> directions, see fig. 1) we have used an average method. More specifically we have run cylindrical oxidation simulations using the <110> and <100> oxidation parameters for the oxidation surface. The final result has been obtained by averaging these two simulations. We have hence used Kao's extension of the Deal and Grove model to 2-D geometry [9], implying both normal stress at the interface and hydrostatic pressure in the oxide volume. A particular attention has been paid in limiting the number of fitting parameters and in using realistic constants. In our model, the initial value of the mean radius (a_0) is deduced from experimental data while oxidations constants are taken from existing literature [7,9,15]. When the NW presents a rectangular section, this radius has been defined as half of the mean value of the lateral dimensions. More details on the oxidation model used in this paper can be found in [16].

By considering input values, we find that the oxidation process does not depend on the diffusion coefficient and is only function of the surface reaction rate. The oxidation is therefore reaction-limited as it is the case in planar oxidation for thin oxides and also for spherical Si nanoparticles [7, 8].

We obtain a good agreement between our model and the experimental evolution of the oxide envelope thickness upon oxidation (see Fig. 2 in dash-dot and dash lines for respectively dry oxidation at 1100°C and wet oxidation at 950°C). In particular, our model perfectly reproduces the retardation of the oxidation rate, so-called "self-limited" oxidation which is characteristics of non planar nanostructures oxidation. This behaviour has already been experimentally observed for nanometric 2D structures but has never been compared to this type of modelling [11-14].

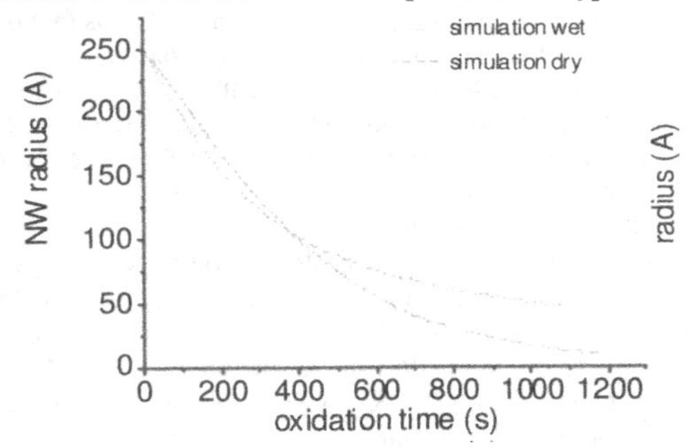

Fig. 3: Theoretical evolution of the NW mean radius (wet oxidation: dash line, dry oxidation: dash-dot line) for a NW with an initial mean radius of 25 nm;

Concerning the NW size evolution, our model predicts that for initial radius of 25 nm, the NW size (see Fig. 3) shrinks down to less than 5 nm in radius for oxidation duration larger than 1000 s in the case of wet oxidation at 950°C and 700 s for a dry process. This size threshold (NW radius < 5 nm) which is targeted for integrating the NW on ultimate transistors, can not be reached directly during the NW fabrication by e-beam lithography due to optical and electronic lithography limitations. This is also the case when fabricating the NWs by other techniques as Vapour Liquid Solid Chemical Vapour Deposition (VLS-CVD).

CONCLUSIONS

Thermal treatments in O_2 at a temperature above the viscoelastic temperature of SiO_2 (950°C) appears to be a powerful way for first controlling the NW size and reduce its diameter down to 10 nm threshold, that could not be reached directly with the state-of-the-art fabrication techniques as e-beam lithography or VLS-CVD. The oxidation rate of the NCs slows down with time and stabilizes after some minutes annealing, thus indicating a self-limiting oxidation process. The Si NWs self-limited oxidation is reaction limited and can be predicted by means of an extended spherical Deal-Grove model taking into account a stress-induced retardation process.

ACKNOWLEDGMENTS

This work is partly financed by the ANR project "PREAANS" (ANR-06-NANO-070).

REFERENCES

1. N. Sing, A. Argawal, L. K. Bera, T. Y. Liow, R. Yang, S. C. Rustagi, C. H. Tung, R. Kurnar, G. Q. Lo, N. Balasubramanian and D. L. Kwong, *IEEE Electron Device Letters* **27/5**, 383 (2006).
2. E. Bernard, T. Ernst, B. Guillaumot, N. Vulliet, X. Garros, V. Maffini-Alvaro, F. Andrieu, V. Barral, F. Allain, A. Toffoli, V. Vidal, V. Delaye, C. vivioz, Y. Campidelli, O. Kemarrec, J. M. Hartmmann, S. Borel, O. Faynot, A. Souifi, P. Coronel, T. Skotnicki and S. Deleonibus, *Proceedings of ESSDERC* 147 (2007).
3. T . Ernst, C. Dupré, C. Isheden, E. Bernard, R. Ritzenthaler, V. Maffini-Alvaro, J. C. Barbé, F. de Crecy, A. Toffoli, C. Vizioz, S. Borel, F. Andrieu, V. Delaye, D. Lafond, G. Rabillé, J. M. Hartmann, M. Rivoire, B. Guillaumot, A. Suhm, P. Rivallin, O. Faynot, G. Ghibaudo and S. Deleonibus, *IEDM Tech Dig.*, 995 (2006).
4. C. Dupré, T. Ernst,, V. Maffini-Alvaro, V. Delaye, J. M. Hartmann, S. Borel, C. Vizioz, . Faynot, G. Ghibaudo and S. Deleonibus, *Solid-State Electronics* **52,** 4, 519 (2008).
5. N. Singh, F. Y. Lim, W. W. Fang, S. C. Rustagi, L. K. Bera, A. Argawal, C. H. Tung, K. M. Hoe, S. R. Omampuliyur, D. Tripathi, A. O. Adeyeye, G. Q. Lo, N. Balasubramanian and D. L. Kwong, *IEDM Tech Dig.*, 548 (2006).
6. S. C. Rustagi, N. Singh, W. W. Fang, K. D. Buddharaju, S. R. Omampuliyur, S. H. G. teo, C. H. Tung, G. Q. Lo, N. Balasubramanian and D. L. Kwong, *IEEE Electron Device Letters* **28/11**, 1021 (2007).
7. H. Coffin, C. Bonafos, S. Schamm, N. Cherkashin, G. Ben Assayag and A. Claverie, M. Respaud, P. Dimitrakis, P. Normand, *J. of Appl. Phys.* **99**, 044302 (2006).
8. B. E. Deal and A. S. Grove, *J. Appl. Phys.* **36**, 3770 (1965).

9. D. B. Kao, J. P. McVittie, W. D. Nix, and K. C. Saraswat, *IEEE Trans. Electron Devices* **35**, 25 (1988).
10. A. Hubert, J. P. Colonna, S. bécu, C. Dupré, V. Maffini-Alavaro, J. M. Hartmann, S. Pauliac, C. Vivioz, F. Aussenac, C. Carabasse, V. Delaye, T. Ernst, S. Deleonibus, *ECS Transactions* **13**, 195 (2008).
11. H. Fukuda, J. L. Hoyt, M. A. McCord, R. F. W. Pease, *Appl. Phys. Lett.* 70 333 (1997).
12. H. I. Liu, D. K. Biegelsen, F. A. Ponce, N. M. Johnson and R. F. W. Pease, *Appl. Phys. Lett.* **64**, 1383 (1994).
13. D. Shi, B. Z. Liu, A. M. Mohammed, K. K. Lew and S. E. Mohney, *J. Vac. Technol.* **B24** 1333 (2006).
14. C. C. Buttner, M. Zacharias, *Appl. Phys. Lett.* **89**, 263106 (2006).
15. A. Fargeix and G. Ghibaudo, *Appl. Phys.* **54**, 7153 (1983).
16. Article submitted to *Appl. Phys. Lett.*

Mater. Res. Soc. Symp. Proc. Vol. 1144 © 2009 Materials Research Society 1144-LL18-15

Nanorods as a precursor for high quality GaN layers

D Cherns[1], I Griffiths[1], S Khongphetsak[1], SV Novikov[2], NRS Farley[2], RP Campion[2], and CT Foxon[2]

[1] Department of Physics, University of Bristol, Tyndall Avenue, Bristol, UK

[2]Department of Physics and Astronomy, University of Nottingham, Nottingham, UK

ABSTRACT

The density of threading dislocations in GaN/(0001)sapphire films grown by molecular beam epitaxy can be reduced to about 10^8 cm^{-2} by growing an intermediate nanorod layer. This paper examines the growth of the nanorods and proposes that threading defects in the overlayer arise either through grain boundaries formed when nanorods coalesce, or through the propagation of dislocation dipoles seen during nanorod growth. Results showing that the latter often terminate or develop into voids during growth are discussed.

INTRODUCTION

The efficiencies of both light emitting and electronic devices based on GaN are often limited by high densities of threading dislocations, generated by the use of highly mismatched substrates. In our work, we have investigated whether the densities of threading dislocations in GaN/(0001)sapphire films, where the lattice mismatch is about 16% and dislocation densities are often in the range $10^9 - 10^{11}$ cm^{-2}, can be reduced by growing an intermediate layer composed of GaN nanorods.

Films have been grown by plasma-assisted molecular beam epitaxy (MBE). As has been known for well over a decade, a transition in MBE-growth of GaN from a relatively smooth two-dimensional layer-by-layer growth mode to a three-dimensional island growth mode can be effected by increasing the overpressure of the nitrogen species, i.e. moving from so-called Ga-rich to N-rich conditions . In fact, the growth of well-defined nanorods under a high overpressure of nitrogen has been extensively studied (see [1] for a review). These studies have shown that, in contrast to many examples of nanorod growth, the MBE growth of GaN nanorods is probably catalyst-free [2]. However, the exact morphology can be strongly influenced by surface preparation, in particular, by the initial deposition of a thin AlN buffer layer [3].

In previous work, we have reported that continuous GaN/(0001)sapphire films with intermediate layers of GaN nanorods can be grown by MBE in a single growth run, and that threading dislocation densities were in the range10^8-10^9cm^{-2} in the overlayer [4]. In this paper, we examine the origin of these threading dislocations in more detail.

EXPERIMENTAL

GaN was grown on (0001)sapphire using plasma-assisted MBE in a Varian ModGen II system. Following the growth of 2-10nm AlN at a substrate temperature of about 600°C, GaN was grown for up to 6hrs at temperatures of 700 - 800°C under a high pressure of activated nitrogen using an HD25 RF-activated plasma source. The nitrogen pressure was then reduced, and growth was continued for further periods of up to 6hrs.

Transmission electron microscopy (TEM) was carried out on samples prepared in cross-sectional orientation by mechanical polishing followed by Ar ion thinning in a Gatan PIPS. The samples were then examined either in a JEOL 2010 TEM operating at 200kV, or in a Philips EM430 TEM operating at 250kV.

RESULTS AND INTERPRETATION

Figure 1 compares a film grown solely in N-rich conditions (Figure 1a) with one grown in N-rich followed by Ga-rich conditions (Figure 1b). Figure 1a shows that the film grown solely under N-rich conditions has a bimodal microstructure consisting of well-defined c-oriented nanorods, emerging from a rough intermediate layer with inclined facets. In fact, a detailed study revealed that the nanorods are Ga-polar, while the intermediate layer is N-polar [4]. The contrast in the nanorods is mainly due to bending, which causes bend contours to run across the nanorods near-perpendicular to the growth direction.

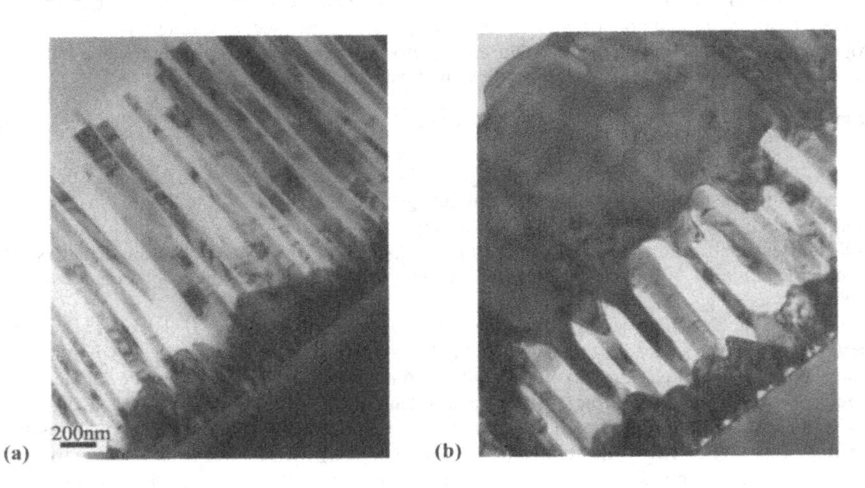

Figure 1: Cross-sectional TEM images of GaN/ 5nmAlN/ (0001)sapphire films (a) 6hrs of GaN growth under N-rich conditions, (b) 5hrs N-rich growth followed by 5hrs Ga-rich growth

In the film which has been subjected to N-rich followed by Ga-rich growth (Figure 1b), it can be inferred that, in the Ga-rich growth, the nanorods have grown laterally rather than vertically to form a relatively continuous overlayer. Growth is predominantly towards the top of the film with little apparent growth towards the base of the nanorods or in the intermediate layer. The overlayer was found to be Ga-polar, i.e. the same polarity as that of the nanorods.

Defects due to nanorod coalescence

A close look at the structure of nanorods in Figure 1b suggests that few, if any, threading defects are generated during the early stages of lateral growth. This can also be seen in Figure 2, where the sample has been oriented to reveal broad-scale differences in contrast across the overlayer due to slight misorientations. This reveals boundaries, such as the one arrowed, which appear to be the meeting front between nanorods which have grown laterally. Indeed, the full extent of lateral growth may be indicated by the in-plane stacking faults which can be seen in the central region of the micrograph. It can be seen that the vertical boundaries consist of threading dislocations. Analyses of these threading dislocations have suggested that most are of perfect type with Burgers vectors of ***a***, ***c***, or ***c+a***. In fact, with ***g*** = 11-20, the dislocations visible in Figure 2 should have ***a*** or ***c+a*** Burgers vectors, with the ***c***-type dislocations being out-of-contrast. We can infer that the threading dislocations are grain boundary dislocations formed when misoriented nanorods coalesce. For example, arrays of threading ***a***-type dislocations would describe a low angle grain boundary due to a relative rotation about [0001].

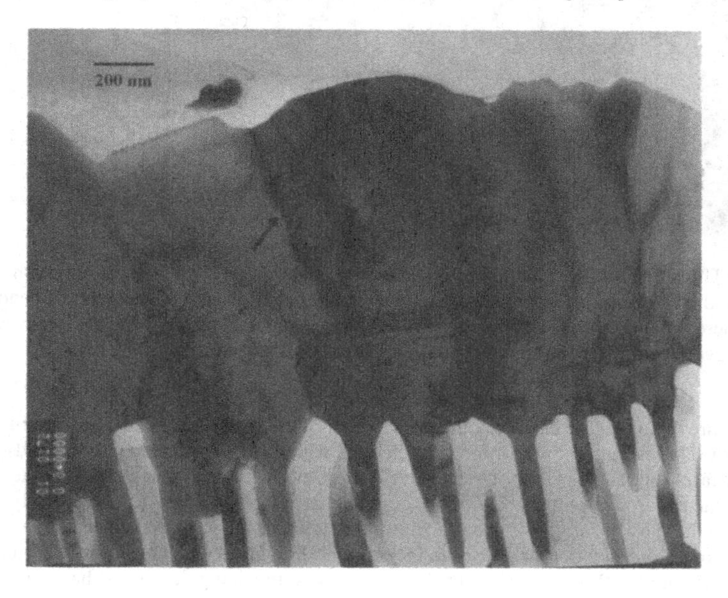

Figure 2: GaN overlayer oriented to give strong diffraction into ***g*** = 11-20, revealing grain boundaries formed where lateral growth fronts meet. These boundaries are seen to have threading dislocations (e.g. see region arrowed).

Defects in nanorods

In general, it has been supposed that nanorods are mostly free of threading defects. However, in recent work, we have shown that the GaN nanorods in our samples frequently contain threading defects [5]. Figure 3 shows an example of these defects, which are planar, generally on {10-10} planes, and bounded by a pair of line defects which show dislocation-like contrast. A detailed analysis has suggested that the bounding dislocations form a partial dipole with Burgers vectors $\boldsymbol{b} = \pm\boldsymbol{c}/2$. Observations of these partial dipoles in a large range of nanorods have suggested some variability: we have noted, for example, that the stacking fault in some cases is replaced by a voided region [5].

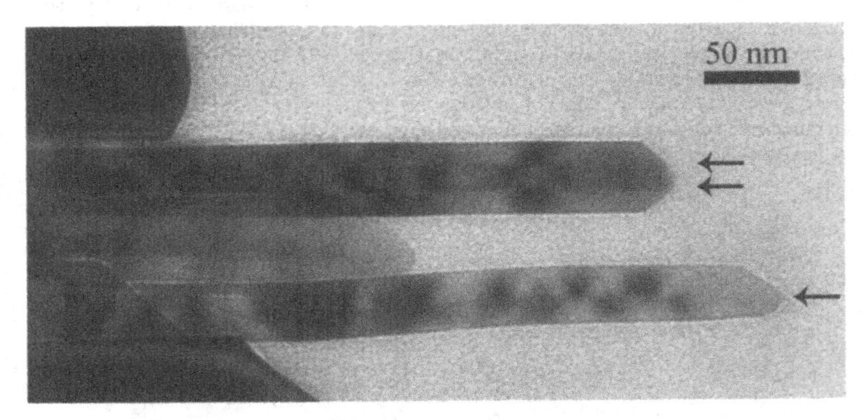

Figure 3: Threading dipoles visible in some nanorods. The single arrow for the bottom nanorod indicates a dipole seen edge on, its presence revealed principally by displacements of bend contours across the defect. The pair of arrows against the top nanorod indicate the two partial screw dislocations forming an obliquely inclined dipole. Displacements of the bend contours, which are of 0002-type, can again be seen.

Figure 4 looks more closely at the propagation of these dipoles during subsequent Ga-rich growth. This sample shows nanorods, such as B, which have grown laterally towards the growth surface (top left), but without extensive coalescence. It is seen that a threading defect present in the nanorod B propagates right through the lateral growth region A, eventually emerging at the growth surface. Figure 4b shows that towards the growth surface, the threading defect becomes void-like in character (n.b. the points A in Figures 4a,b correspond). Figure 4c shows the region near B in Figure 4a, illustrating a second threading defect that has terminated during growth.

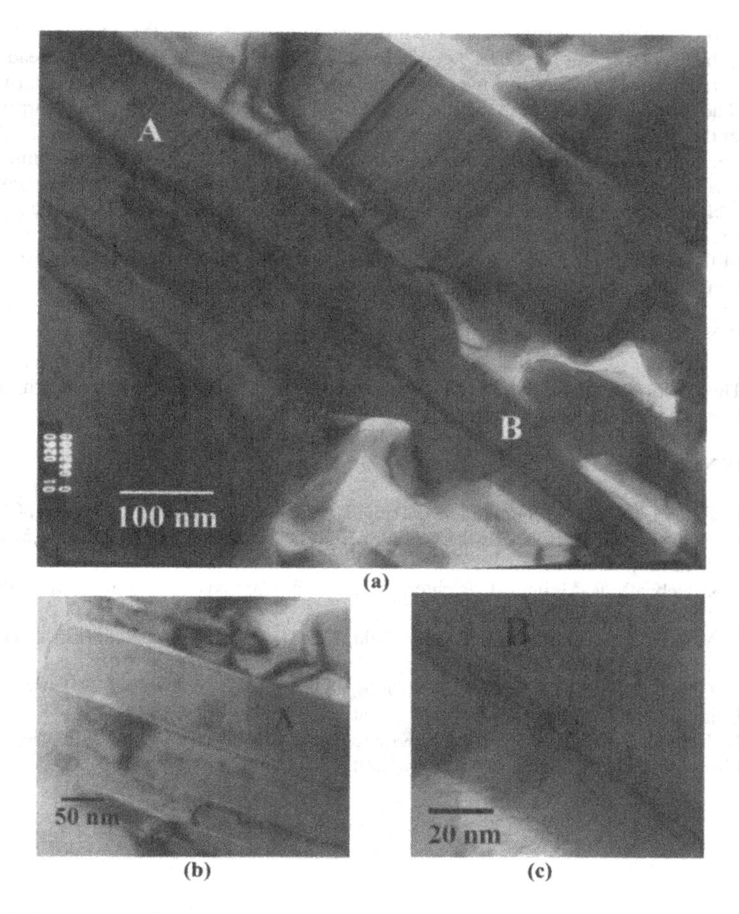

Figure 4: (a) A nanorod B which has grown laterally during Ga-rich growth without joining to neighbouring nanorods. Of two threading dipoles in the original nanorod, one has terminated during growth, near the point B (see blow-up of this region in (c)). Towards the growth surface, the second threading dipole has changed character to display a voided region, seen more clearly in (b).

DISCUSSION

The results in Figures 2-4 suggest that, as nanorods grow laterally and eventually coalesce, few new threading dislocations arise before coalescence. This implies that threading dislocations in the overlayer arise in only one of two ways, either (a) in the boundaries formed where nanorods coalesce, or (b) through the propagation of pre-existing threading defects in the nanorods themselves. Figure 2 shows a good example of the former. In the case of threading dipoles, there is good evidence that many such defects change structure or terminate during growth. The formation of a void in the threading dipole seen in Figure 4b is not too surprising, given that the displacement of $\boldsymbol{b} = c/2$ defines a stacking fault that is probably relatively high energy, i.e. with a stacking sequence changing from ABABAB to BABABA. The termination of the faults is also readily understandable, given that the bounding partials may simply recombine to form a closed loop. The result is that we expect most of the threading defects in thick continuous GaN overlayers to arise through nanorod misorientations. In this case, the overall density of threading dislocations can be expected to depend on the relative alignment of nanorods, and on the extent of lateral overgrowth, i.e. on the nanorod density.

ACKNOWLEDGEMENTS

The authors are grateful to the UK Engineering and Physical Sciences Research Council (EPSRC) for support for this work under grant EP/D080762/1.

REFERENCES

1. E. Calleja, J. Ristic, S. Fernandez-Garrido, L. Cerruti, M.A. Sanchez-Garcia, J. Grandal, A. Trampert, U. Jahn, G. Sanchez, A. Griol and B. Sanchez, Phys. Stat. Sol. b **244,** 2816-2837 (2007)
2. R.K. Debnath, R. Meijers, T. Richter, T. Stoica, R. Calarco and H. Lueth, Appl. Phys. Lett. **90,** 123117 (2007)
3. K.A. Bertness, A. Roshko, L.M. Mansfield, T.E. Harvey and N.A. Sanford, J. Cryst. Growth **300,** 94-97 (2007)
4. D. Cherns, L. Meshi, I. Griffiths, S. Khongphetsak, S.V. Novikov, N.R.S. Farley, R.P. Campion and C.T. Foxon, Appl. Phys. Lett. **92,** 121902-4 (2008)
5. D. Cherns, L. Meshi, I. Griffiths, S. Khongphetsak, S.V. Novikov, N.R.S. Farley, R.P. Campion and C.T. Foxon, Appl. Phys. Lett. 93, 111911-3 (2008)

Mater. Res. Soc. Symp. Proc. Vol. 1144 © 2009 Materials Research Society **1144-LL18-22**

Stabilizing Dispersions of Large Quantities of Selenium Nanowires

Michael C.P. Wang and Byron D. Gates
Simon Fraser University, 8888 University Drive, Burnaby, B.C. Canada V5A 1S6

ABSTRACT

Through solution-phase synthesis we can produce large quantities of high aspect-ratio selenium nanowires. The dispersion and assembly of these nanowires is, however, a challenge due to aggregation and settling of the material from solution. It is desirable to develop techniques for the incorporation of these nanostructures as key components in advanced materials and applications taking advantage of their properties. We have advanced the understanding of how to control the dispersion of selenium nanowires. This control is a key step in directing the assembly of the nanostructures.

INTRODUCTION

Selenium nanowires are synthesized in large quantities by solution-phase techniques. Following this synthesis the nanowires are entangled and quickly settle out of solution. Aggregation of selenium nanowires is most likely driven by the favorable interaction of their exposed, high-energy surfaces (Figure 1). Although the nanowires can be dispersed through serial dilution, processing large quantities of nanowires by this technique is impractical (Figure 1a). In addition, techniques that are used to disperse other nanowires or wire-like materials (e.g., sonication)[1-3] creates a uniformly red dispersion in solution (Figure 1c). However, upon closer inspection the integrity of the selenium nanowires is not maintained, and the sample is composed of nanorods. New methods of dispersion are required for processing large quantities of selenium nanowires.

A universal approach to uniformly disperse the nanowires is to cap or otherwise mask the surfaces of these nanostructures through interactions with a solvent and/or surfactant. Investigating the favorability of a surfactant to interact with the surfaces of nanowires is essential to building an understanding of appropriate techniques for stabilizing dispersions of these nanowires. Properties of these nanowires could be extended to stabilizing the suspension of other selenium nanostructures in various solvents. This challenge is important for subsequent reactions on these nanostructures. Selenium has been previously demonstrated to be an important template for the formation of other nanostructured materials.[4, 5] Overcoming the tendency of selenium nanostructures to settle from solution is also important for the subsequent assembly of these materials into well-defined patterns. Organizing these materials through an assembly process is essential for taking advantage of their unique optoelectronic properties.[6]

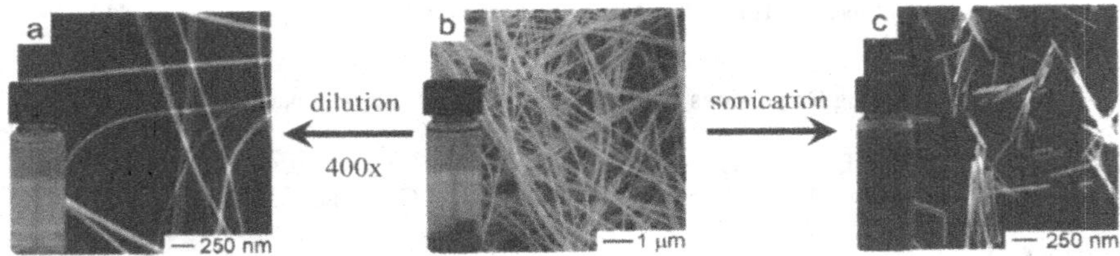

Figure 1. Optical images of selenium nanowires dispersed in an ethanol solution, and scanning electron microscopy (SEM) images of nanowires dispersed onto silicon substrates indicate the relative concentration and morphology of the nanostructures, respectively. These solutions have been processed by (a) dilution from (b) the original concentration, or (c) agitation of the original solution by sonication.

We have investigated the interaction of a number of solvents and surfactants to build an understanding of how to stabilize suspensions of selenium nanowires. In this study, we demonstrate the capability of dispersing and stabilizing selenium nanowires as suspensions in various solvents. We compare the results using both quantitative and qualitative measurements. From our experimental results we confirmed that the surfaces of selenium nanowires, when synthesized through the sonochemical approach,[7, 8] are hydrophobic, and with the appropriate surface modification these nanowires can be dispersed and stabilized in both organic and aqueous environments.

EXPERIMENTAL

The synthesis of trigonal selenium (*t*-Se) nanowires requires an initial synthesis of amorphous selenium (*a*-Se). This precursor material was prepared by modifying a previously reported procedure.[7] To prepare amorphous selenium, a solution containing 2.73 g (21.1 mmol) of selenious acid (H_2SeO_3; 98%; Sigma-Alrich), was dissolved in 100 mL of 18.0 MΩ water (Millipore). The mixture was prepared in a 250 mL round bottom flask containing a 1-cm long magnetic stir bar. Hydrazine hydrate ($N_2H_4 \cdot xH_2O$; 50-60%; Sigma-Aldrich), 3 mL (61.1 mmol), was transferred into a 5 mL glass vial. These two solutions were chilled in an ice-water bath for 15 min. The hydrazine was subsequently added drop-wise over a period of 2 min to the vigorously stirring solution of selenious acid while this solution is kept in the ice-water bath. The mixture slowly changed from an initial pale orange to a brick-red color over a period of 10 min. This red solution contains suspended *a*-Se particulates. The red suspension was filtered through a 0.1 μm polycarbonate (PC) membrane (Millipore) and the filtrate was washed with 200 mL of high purity water. The filtrate was dried in a desiccator in the absence of light for 2 days prior to using this precursor in the synthesis of the *t*-Se nanowires.

To synthesize *t*-Se nanowires, a 2 mg sample of *a*-Se was dispersed in 1 mL of isopropanol (*i*PA; ≥99.5%; Anachemia) in a glass vial. This solution was sonicated for 20 s in a water bath at room temperature to yield a red-colored homogeneous solution that was stored in darkness at 20 °C for 12 h. During this period of time crystalline selenium nanowires growth in solution consuming the *a*-Se material. The nanowires easily settled from solution. Further processing of the nanowires required manual agitation in addition to dilution of the suspension by 40 times with more *i*PA. Suspensions of the nanowires were portioned into 1.5 mL aliquots and transferred into centrifuge tubes. These solutions were spun at 1000 rpm for 30 min. The solvent was decanted and replaced with another organic solvent of interest (see Discussion for further details), and this purification procedure repeated for a total of 3 times to ensure the removal of the *i*PA. However, when dispersing nanowires in aqueous solutions a surfactant is added to the solution during each solvent exchange. Upon removal of the *i*PA, this mixture of nanowires and surfactant are manually shaken for ~30 s to disperse the nanowires in this aqueous medium.

All chemicals and solvents were used as received.

DISCUSSION

The dispersions of selenium nanowires are qualitatively evaluated using optical images of the suspensions taken with a digital camera. Uniformly dispersed nanowires resulted in a homogeneous red color within the solution, such as nanowires dispersed in ethanol. In contrast, selenium nanowires poorly dispersed in a solvent (e.g., directly suspended in water) produced observable aggregates (Figure 2). In this case, the ethanol serves as a stabilizing surfactant for the nanowires. Decreasing the percent ethanol in the solution to 80 % (v/v in water) alters the appearance of the solution. Granular precipitates are noticeable in the digital photograph although the solution remained red in appearance. At a concentration of 50 % ethanol/50 % water the solution is no longer homogeneous in appearance. The red colored dispersion is replaced with large aggregates of nanowires. In addition, as more water is introduced into the dispersion the number of small aggregates decreases at the expense of the formation of larger aggregates.

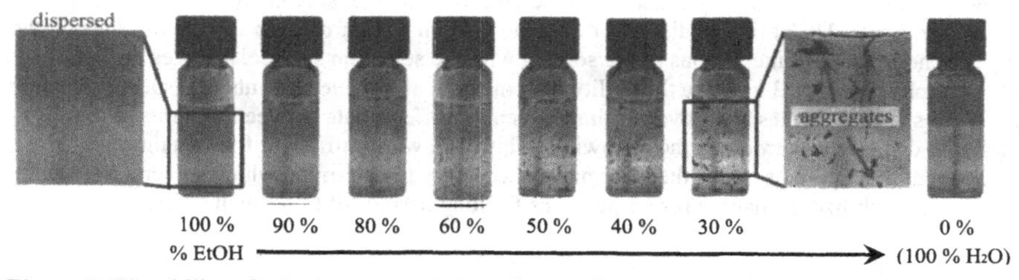

Figure 2. The ability of selenium nanowires to disperse in a solution of ethanol decreases as the water content increases, as indicated by the appearance of aggregates in these optical images.

The trend observed when dispersing selenium nanowires in ethanol with varying amounts of water present suggests that the surface of these nanowires is hydrophobic when synthesized sonochemically. We can take advantage of this property to assist in a establishing a more favorable condition to create uniform dispersions. We hypothesize that through the assistance of surfactant molecules or block copolymers consisting of hydrophilic and hydrophobic portions we would be able to disperse these hydrophobic nanowires in aqueous solutions. The non-polar segment of surfactant should interact with the non-polar surface of selenium nanowires while the hydrophilic regions would be exposed to the aqueous environment.[9] Hence, the nanowires should be stabilized in solution.

We have demonstrated the dispersion of selenium nanowires in water using polymers that favorably interact with the surface of the nanowires while presenting a polar surface to the aqueous environment. The addition of either polyvinyl alcohol (PVA) or Brij 30 to a mixture of water and selenium nanowires can disperse the nanowires. The non-polar backbone of the PVA or the hydrophobic tail of the Brij 30 interacts with the surface of the nanowires (Scheme 1).

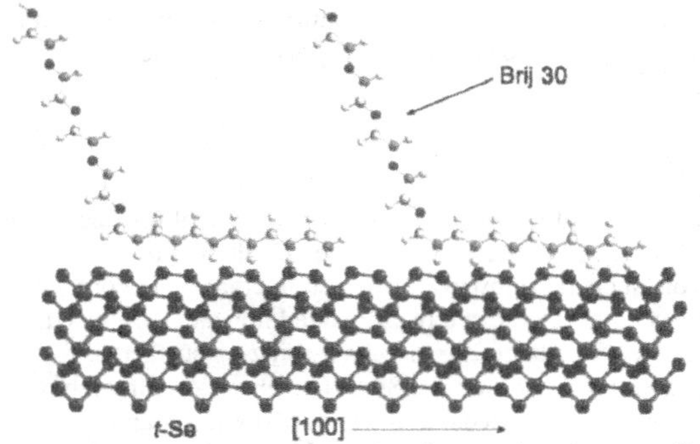

Scheme 1. Proposed non-covalent interactions between Brij 30 molecules and the surface of *t*-Se nanowires.

The soluble hydroxyl groups and hydrophilic polyethylene glycol segments of each polymer, respectively, are able to disperse the selenium nanowires in water (Figure 3). This interaction between the selenium and the surfactant can be extended to other solvent systems, such as polar organic solvents. The ability to disperse *t*-Se nanowires in a polar organic solvent depends on the types of non-bonding interactions of the solvent with the selenium surface. A series of digital photographs in Figure 3 indicate the ability of some typical organic solvents to disperse selenium nanowires. These images are, however, insufficient to differentiate between solvents in their ability to disperse *and stabilize* the nanowires. This data, while sufficient for a qualitative assessment of dispersion, lacks the information necessary to determine which solvent/surfactant system can stabilize the nanowires suspensions for long periods of time, such as hours or days.

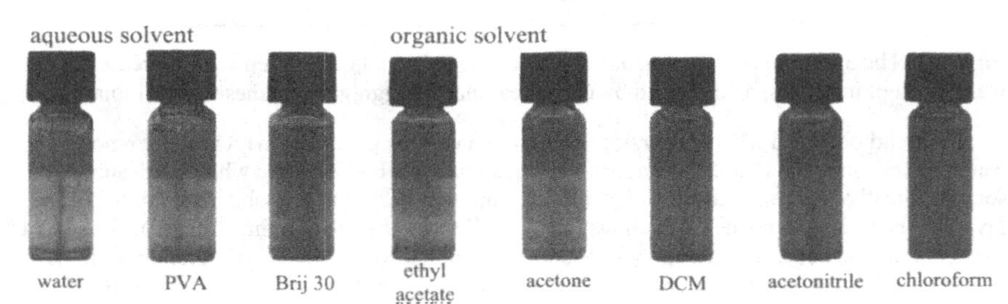

Figure 3. Optical images of selenium nanowires dispersed in organic and aqueous solutions. (PVA = polyvinyl alcohol; DCM = dichloromethane)

We monitored the changes in absorbance for various solutions to determine the ability of various solvent systems and surfactants to keep nanowires dispersed in solution for a period of time. In order to perform this study, we used a UV-visible spectrometer to monitor the changes in absorbance of the red dispersions. We monitored the settling of the selenium nanowires by observing changes in the absorbance of each solution at 400 nm. This spectral position corresponded to an absorbance band of the selenium nanowires. We monitored changes in absorbance at 10 min intervals over a period of 4 h, and averaged these measurements from three independent runs performed on the same sample (Figure 4). All measurements were also carried out on the same batch of nanowires. Any decrease from the normalized initial absorbance intensity at 400 nm indicated settling of the selenium nanowires from solution. Our evaluation of nanowires dispersions in the polar organic solvents (e.g., dichloromethane, chloroform, and acetonitrile) displayed an exponential decrease in the signal intensity. This trend indicates the lack of a favorable interaction between the solvent molecules and the nanowires that is required to stabilize the nanowires suspensions. These molecules have a relatively small or negligible non-polar region that interacts weakly with the surfaces of the nanowires. On the other hand, acetone and ethyl acetate contain methyl and ethyl non-polar moieties, respectively. The non-polar regions within the acetone and ethyl acetate most likely stabilize the solvent interactions with the nanowires. These solvents stabilize the nanowires through favorable non-polar interactions (Figure 4a). We extended this analysis to the nanowires dispersed in aqueous environments. Our studies revealed that the polymer assisted dispersion of the nanowires in the aqueous solutions stabilized the suspensions of nanowires for periods of time greater than 4 h (Figure 4b). The interactions of the polymer with the nanowires are significantly more stable than that of the individual solvent molecules. This result is attributed to the multiple interactions of each polymer with the surfaces of the nanowires. These results suggest that a number of surfactants with a hydrophobic tail of appropriate length can be designed to interact with the selenium nanowires to further stabilize them in different solvent environments.

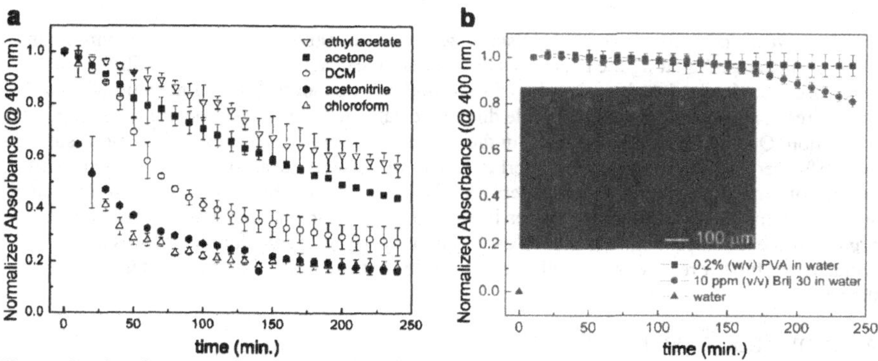

Figure 4. Absorbance measurements (at 400 nm) plotted over a period of 4 h for suspensions of selenium nanowires in various (a) organic and (b) aqueous solutions. Inset of (b) shows an optical image of nanowires drop-cast onto a silicon wafer from an aqueous dispersion in Brij 30.

The previously discussed quantitative measurements are only meaningful when comparing selenium nanowires produced from the same initial nanowires synthesis. Varying the length of the nanowires can drastically change our ability to stabilize dispersions of these nanostructures. In order to demonstrate the importance of aspect ratio of dispersion of the selenium nanostructures we compared the dispersion of nanorods (Figure 1c) and nanowires (Figure 1b). Nanorods are more easily stabilized in suspension than the nanowires (Figure 5) when comparing the nanostructures dispersed in an identical solvent system. The nanowires are inadequately stabilized in the methanol, but the nanorods slowly settle from methanol. A solution of *i*PA stabilizes the nanorods. The observed trend is comparable to that for the nanowires dispersed in an aqueous solution containing PVA. The nanowires are, however, not stabilized and settle out of solution over time.

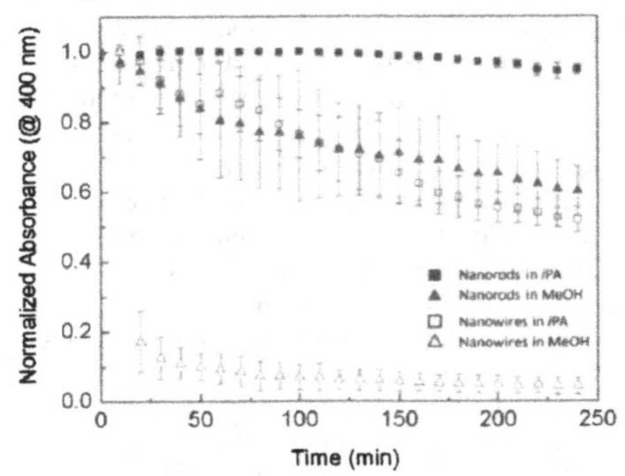

Figure 5. A settling study comparing selenium nanowires and selenium nanorods dispersed in isopropanol and methanol over a period of 240 min.

CONCLUSIONS

We have successfully dispersed and stabilized selenium nanowires in aqueous environments using surfactants. Agitating the nanowires by sonication can also increase the stability of the nanowires in solution, but this is attributed to formation of nanorods by fragmentation of the nanowires. This procedure is unfavorable due to the damage to the nanostructure following sonication. Quantitative measurements on the stability of nanowires dispersed in various solvents are performed by monitoring light scattering and absorbance of the suspensions over extended periods of time. Surfactant assisted dispersion of selenium nanowires in aqueous solutions has a superior ability to stabilize the nanowires in comparison to dispersion in a number of polar organic solvents. These findings will be essential to utilizing the selenium nanowires as templates in further reactions, as well as to assemble these materials into new materials and devices.

ACKNOWLEDGEMENTS

This research was supported in part by the Natural Sciences and Engineering Research Council (NSERC) of Canada, the Canada Research Chairs Program, and Simon Fraser University (SFU). This work made use of 4D LABS shared facilities supported by the Canada Foundation for Innovation (CFI), British Columbia Knowledge Development Fund (BCKDF), and SFU

REFERENCES

1. J. R. Yu, N. Grossiord, C. E. Koning, J. Loos, *Carbon* **45**, 618 (2007).
2. B. R. Priya, H. J. Byrne, *J. Phys. Chem. C* **112**, 332 (2008).
3. M. Zheng, A. Jagota, E. D. Semke, B. A. Diner, R.S. Mclean, S. R. Lustig, R. E. Richardson , N. G. Tassi, *Nature Mater.* **2**, 338 (2003).
4. B. Gates, Y. Wu, Y. Yin, P. Yang, Y. Xia, *J. Am. Chem. Soc.* **123**, 11500 (2001).
5. B. Gates, B. Mayers, Y. Wu, Y. Sun, B. Cattle, P. Yang, Y. Xia, *Adv. Funct. Mater.* **12**, 679 (2002).
6. J. C. Chou, S. Y. Yang, Y.S. Wang, *Mater. Chem. Phys.* **78**, 666 (2003).
7. B. D. Gates, B. Mayers, A. Grossman, Y. Xia, *Adv. Mater.* **14**, 1749 (2002).
8. B. T. Mayers, K. Liu, D. Sunderland, Y. Xia, *Chem. Mater.* **15**, 3852 (2003).
9. L. Vaisman, H. D. Wagner, G. Marom, *Adv. Colloid and Interface Sci* **128-130**, 37 (2006).

Mater. Res. Soc. Symp. Proc. Vol. 1144 © 2009 Materials Research Society 1144-LL18-25

Grafting of Organic Molecules on Silicon Nanowires

Kaoru Kajiwara[1], Masato Ara[2] and Hirokazu Tada[1]
[1] Division of Materials Physics, Graduate School of Engineering Science, Osaka University, Toyonaka, 560-8531, Japan.
[2] Organization for the Promotion of Research on Nanoscience and Nanotechnology, Osaka University, Toyonaka, 560-8531, Japan.

ABSTRACT

We have studied the surface modifications of silicon nanowires (SiNWs) by grafting of organic molecules *via* Si-C covalent bonds. SiNWs were prepared by thermal evaporation of SiO powders under the H_2 / Ar atmosphere. Transmission electron microscope observation revealed that SiNWs with a diameter of 50 ~ 100 nm were synthesized and consisted of the single-crystalline Si core with Si oxide sheath. Perfluoro-octylethylene was covalently anchored to the surface of SiNWs by wet process and electrical properties of SiNWs were measured. Modified SiNWs showed a p-type semiconducting behavior. It was presumed that this effect was caused by the charge transfer between the SiNWs and the molecules. The surface modification with organic molecules is a useful method for controlling electronic characteristics in SiNWs.

INTRODUCTION

Silicon nanowires (SiNWs) have received considerable attention because of their potential application to building blocks of nano- and micro- electronics devices. The surface of SiNWs plays a major role for solid-state properties since the surface-to-volume ratio is very high. Therefore, surface modifications are promising ways to control the solid-state properties of SiNWs. We have examined the surface modification by using organic monolayers. Several techniques have been established to graft molecules onto Si surfaces *via* Si-O-C and Si-C covalent bonds [2,3]. Grafting of molecules *via* Si-C covalent bonds is useful because they are very stable thermally and chemically [3]. In addition, it can control the solid-state properties directly because there is no intervening oxide layer between molecules and the surface of SiNWs.

In the present study, we investigated the electrical properties of SiNWs modified with perfluoro-octylethylene having a large electron affinity.

EXPERIMENT

Synthesis of SiNWs

SiNWs were prepared by thermal evaporation of SiO powders (Aldrich, -325 mesh) [4]. The SiO powder was placed into an alumina crucible and degassed at approximately 900 K under vacuum. After the sufficient degassing, it was annealed at 1200 K for 10 minutes under the H_2 (5 %) / Ar atmosphere with a pressure of approximately 10 Torr. SiNWs were produced in the crucible. The products were taken from the crucible and suspended in 2-propanol. The

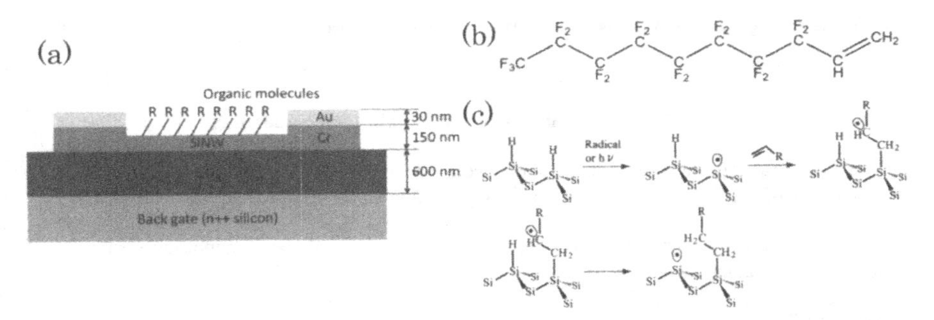

Figure 1. (a) The schematic view of the devices. (b) The molecular structure of perfluoro-octylethylene and (c) the schematic of the reaction between an alkene and H-Si surface.

suspension was spread on a transmission electron microscope (TEM) grid for TEM observation (HITACHI, H-800) and on a silicon plate for atomic force microscope (AFM) observation.

Device Fabrication

Fig. 1 (a) shows the schematic view of the device for two-terminal, gate-dependant transport measurements of SiNWs. At first, synthesized SiNWs were treated by a 5 % HF solution for 1 h to remove the oxide sheath and suspended in 2-propanol. Then, suspended SiNWs were dropped on Si substrates (ca. 0.02 Ω cm) with 600 nm SiO_2 as an insulator. Electrical contacts to the isolated SiNW were made by using standard electron beam lithography and liftoff processes. Thin films of Cr (150 nm) and Au (30 nm) were deposited by thermal evaporation. The gap length was 2 μm. Au / Cr electrodes were chosen because of the high stability against HF and NH_4F solution. Thermal annealing was carried out at 1073 K for 1 h under vacuum to improve the contact between Cr and the surface of SiNWs.

Device Characterization and Molecular Grafting

The devices were introduced to a probe station with a base pressure of 5×10^{-3} Pa. The metal tips were placed onto the source / drain electrodes *via* Ag paste. First, the devices were tested before the molecular grafting. After the measurement, the devices were transferred into a glovebox in which grafting of molecules on SiNWs was performed. In this study, perfluoro-octylethylene was grafted on the surface of SiNWs. **Fig. 1** (b) and **1** (c) show the structure of perfluoro-octylethylene and the schematic of the reaction of an alkene on H-Si, respectively. The hydrogen termination of SiNWs on the devices carried out by the treatment in a HF solution (1 %) for 30 sec and a NH_4F solution (10 %) for 1 min. The devices were put into neat perfluoro-octylethylene deoxygenated and heated at 423 K for 5 h on a hotplate. Then, devices were rinsed in petroleum ether, ethanol and dichloromethane. After the grafting process, electrical measurement was carried out again.

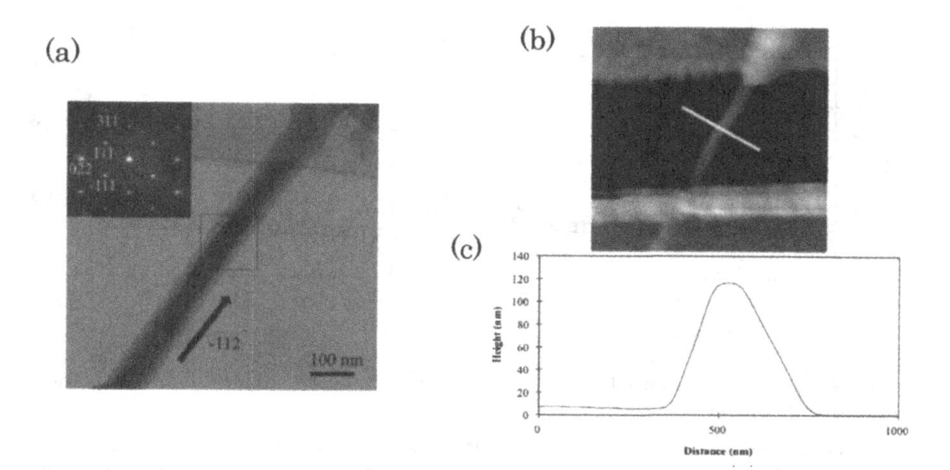

Figure 2. (a) TEM image of SiNW synthesized from SiO powder. The wire consisted of a Si crystalline core and Si oxide sheath. Inset is the diffraction pattern taken at the square area. (b) A topographic AFM image (2 μm × 2 μm) of a SiNW and (c) the height profile.

RESULTS & DISCUSSION

Fig. 2 (a) shows a TEM image of the SiNW synthesized from the SiO powder. The diameter of the SiNW was approximately 100 nm. The length was a few μm. Continuous gradation of contrast was observed at the cores of SiNWs. On the other hand, sheaths of SiNWs were observed as amorphous. It is known that the sheaths of SiNWs prepared from SiO consisted of silicon oxide [4]. The thickness of oxide sheath was approximately 10 nm and it was thicker than that of native oxide. The inset is the diffraction pattern taken at the square area. It shows the core of SiNW consists of single-crystalline silicon and SiNW grew along the [-1 1 2] direction. This is consistent with the previous results [5]. **Fig. 2** (b) and **2** (c) show an AFM image of a SiNW and its cross section view, respectively. The height of SiNWs was 117 nm, which agreed well with the diameter obtained by TEM.

Fig. 3 shows the characteristics of the device before and after the grafting of perfluorooctylethylene. The source-drain bias (V_{SD}) was swept from -30 V to 30 V. The gate bias (V_G) was swept from -80 V to 80 V. The source-drain current (I_{SD}) of SiNWs without molecules grafted was very low and did not depend on the gate voltage (V_G). On the other hand, after the grafting of molecules, I_{SD} increased with V_G biased negatively. This result indicated that this SiNW showed the p-type semiconducting behavior.

This behavior was thought to be explained as follows. Since perfluoro-octylethylene exhibits large electron affinity, it acts as an electron acceptor. When the molecules are anchored to SiNWs, they tend to withdraw electrons from SiNWs. SiNWs are thus expected to be positively charged slightly due to the increase of hole density, which results in the shift of Fermi level towards the valence band. Then the barrier height between electrodes and SiNWs is reduced for smooth injection of holes from the electrodes to SiNWs.

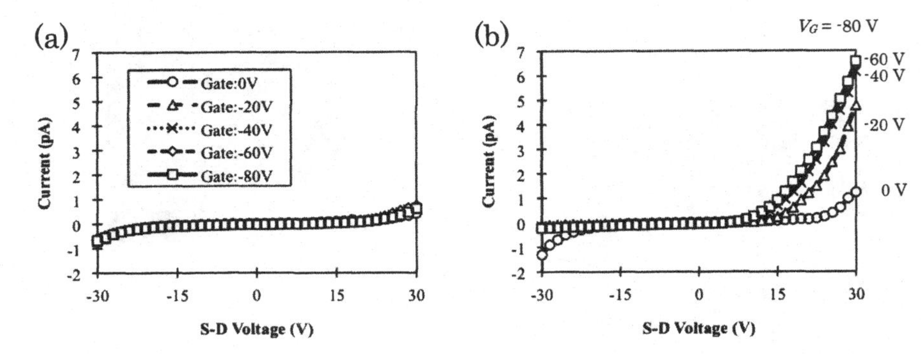

Figure 3. The I-V characteristic of SiNWs with native oxide sheath (a) and organic monolayer (b).

On the assumption that the surface of SiNWs with a diameter of, for example, 50 nm, is covered with densely packed molecules, the total number of molecules grafted is estimated to be 8.0×10 cm^{-2} [6]. If each molecule is assumed to be charged negatively at $10^{-3} \sim 10^{-9}$e, the hole density in SiNWs is estimated to be approximately $10^{18} \sim 10^{12}$ cm^{-3}. The Fermi level, then, shifts by 0.5 eV ~ 0.1 eV towards the valence bands.

CONCLUSIONS

In conclusion, we demonstrated the synthesis of SiNWs by thermal evaporation. SiNWs with a diameter of 100 nm were synthesized. SiNWs were found to consist of single-crystalline Si core and oxide sheath. We grafted perfluoro-octylethylene on the surface of SiNWs. It was found that perfluoro-octylethylene grafted SiNWs showed the p-type semiconducting behavior. This behavior is probably due to the charge transfer from SiNWs to molecules having large electron affinity. The surface modification of SiNWs with organic monolayers is found to be a promising way to tune the electrical properties of SiNWs.

ACKNOWLEDGMENT

A part of the present experiment was carried out by using a facility in the Research Center for Ultrahigh Voltage Electron Microscopy, Osaka University.

REFERENCES

1. W. Lu and C. M. Lieber, *J. Phys. D.* **39**, R387-R406 (2006).
2. J.Sagiv, *J. Am. Chem. Soc.* **102**, 92 (1980).
3. M. R. Linford, P. Fenter, P. M. Eisenberger and C. E. D. Chidsey, *J. Am. Chem. Soc.* **117**, 3145 (1995).

4. W. S. Shi, H. Y. Peng, Y. F. Zheng, N. G. Shang, Z. W. Pan, C. S. Lee and S. T.Lee *Adv. Mater.* **12**, 1343 (2000).
5. N. Wang, Y. F. Zhang, Y. H. Tang, C. S. Lee and S. T. Lee, *Appl. Phys. Lett.* **73**, 3902 (1998).
6. A. B. Sieval, R. Linke, H. Zuilhof and E. J. R. Sudholter, *Adv. Mater.* **12**, 1457 (2000).

Mater. Res. Soc. Symp. Proc. Vol. 1144 © 2009 Materials Research Society 1144-LL19-06

Electronic Structure and Magnetization of Diluted Magnetic Semiconductor Nanowires

Yong Jae Cho, Kyung Hwan Ji, Chang Hyun Kim, Han Sung Kim, Yong Jei Son, and Jeunghee Park*

Department of Materials Chemistry, Korea University, Jochiwon, 339-700, Korea, Republic of

ABSTRACT

$Ga_{1-x}Mn_xAs$ ($0 < x < 0.05$) and $Ti_{1-x}Mn_xO$ ($0 < x < 0.03$) nanowires were synthesized by thermal vapor transport method. They consisted of single-crystalline nanocrystals, with uniform grown direction. The lattice constant decreases initially with increasing amount of doping content (x), and then increases as x increases. X-ray photoelectron spectroscopy revealed that as the Mn content increases, the binding energy of Ti 2p shifts to a higher energy, suggesting the possibility of hybridization between the Mn^{2+} ions and host defects. X-ray absorption spectroscopy and X-ray magnetic circular dichroism confirmed that the Mn^{2+} ions substitute into the tetrahedral coordinated sites. The magnetization measurement revealed that all of these nanowires exhibited room-temperature ferromagnetic behaviors.

INTRODUCTION

Diluted magnetic semiconductors (DMSs), in which the host cations are randomly substituted by magnetic ions, have attracted a considerable amount of attention, because of their excellent potential as key materials for spintronic devices.[1-3] The demonstration of the unique phenomena associated with DMSs, such as the field-effect control of their ferromagnetism, efficient spin injection to produce circularly polarized light, and spin-dependent resonant tunneling, has opened up a rich and various landscape for technological innovation in magnetoelectronics. As one of the most important ferromagnetic DMSs, Mn-doped GaAs ($Ga_{1-x}Mn_xAs$) has been extensively studied for more than two decades.[4-16] The spin 5/2 Mn (Mn^{2+}) ions at the regular sites of the zinc blende lattice of the GaAs host act as acceptors, thus providing both magnetic moments and itinerant holes which mediate the ferromagnetic order. Nowadays, their ferromagnetism is fairly well understood, thus allowing their Currie temperatures to be predicted. In this respect, Mn-doped GaAs is one of the best understood ferromagnetic materials. Dietl et al. predicted the Curie temperature (TC) of 5% Mn-doped p-type GaAs ($Ga_{0.95}Mn_{0.05}As$) to be 120 K, by the Zener model description.[6] The quest to increase the TC to room temperature has led to a thorough investigation of the material properties. Experimentally, it was reported that annealing (< 200 Δ) for a long time can lead to a significant improvement in its TC, with values as high as 159 K being reported.[15] The magnetic properties were enhanced with a value of TC = 250 K being reached, as it forms Mn-δ doping heterostructures.[16]

The majority of these studies on $Ga_{1-x}Mn_xAs$ focused on the bulk materials, but its integration into electronics will require very low dimensions in order to make real use of the advantages offered by the spins. The calculations predicted that the quantum confinement effect can even increase the magnetic properties in certain nanosize regimes. Recently, intensive research activities have been directed toward one-dimensional (1D) nanostructures, which are considered to be building blocks for the fabrication of various nanoscale devices. In particular, well-defined single-crystalline nanowires enable us to scrutinize precisely the magnetic properties and electronic structures of DMSs depending on their crystal size and Mn doping levels. There have been a number of reports on the synthesis of Mn (< 20%)-doped GaAs nanowires, showing a wide range of TC values from 30 K to 350 K. In order to fully understand their magnetic properties, it is important to elucidate the electronic structures which are responsible for their magnetic properties. However, the correlations between the Mn content, electronic structures and magnetization are far from being well understood.

Herein, we synthesized high-purity single-crystalline $Ga_{1-x}Mn_xAs$ nanowires (NWs) with controlled Mn contents, x = 0, 0.01, 0.02, 0.03, and 0.05, exhibiting room-temperature ferromagnetism, using a simple vapor transport method. The Mn content was controlled by using different evaporation temperatures of the Mn source ($MnCl_2$). The ferromagnetic behaviors at room temperature (TC > 350 K) were identified for all Mn content. Their lattice constants and electronic structures were thoroughly investigated as a function of the Mn concentration by high-resolution X-ray diffraction (XRD), X-ray photoelectron spectroscopy (XPS), X-ray absorption spectroscopy (XAS), and X-ray magnetic circular dichroism (XMCD). We attempted to correlate the magnetic moments of these nanowires with their electronic structures. In addition High-density TiO_2 NWs were synthesized on a large area of Ti substrates, by the thermal oxidation of Ti foil, combined with vapor transport of Mn. Magnetization measurement revealed that 1% Mn-doped TiO_2 nanobelts exhibit room-temperature ferromagnetic behaviors.

EXPERIMENTAL DETAILS

1. Mn doped GaAs nanowires

GaAs (99.999%, Aldrich) and $MnCl_2$ (99.99%, Aldrich) powders were placed separately in two quartz boats loaded inside a quartz tube reactor. A silicon substrate, on which a 3-5 nm thick Au film was deposited, was positioned at a distance of 10 cm away from the GaAs source. Argon gas was continuously flowed at a rate of 500 sccm during the synthesis. The temperatures of the Ga and Mn sources were set to 1100 and 800 °C, respectively, and that of the substrate was approximately 800-850 °C.

2. Mn doped TiO_2 nanowires

The Mn-doped TiO_2 NWs were synthesized by a two-step process; (1) the preparation of TiO_2 NWs, A piece of Ti foil (10 x 5 x 0.25 mm^3), on which a 3-5 nm thick Au film was deposited, was placed on the top of the quartz boat and water was placed 5cm in front of the quartz, boat located inside a quartz tube reactor, The tube reactor was then heated electrically to 750 °C under Ar gas flow 300 sccm during the reaction time at 1 hour. (2) their conversion to Mn-doped TiO_2 NWs. For the conversion reaction, the as-grown TiO_2 NWs were placed inside reactor tubes, and $MnCl_2$ beads were evaporated at 750 °C for 30 min.

RESULT AND DISCUSSION

1. Mn doped GaAs nanowires

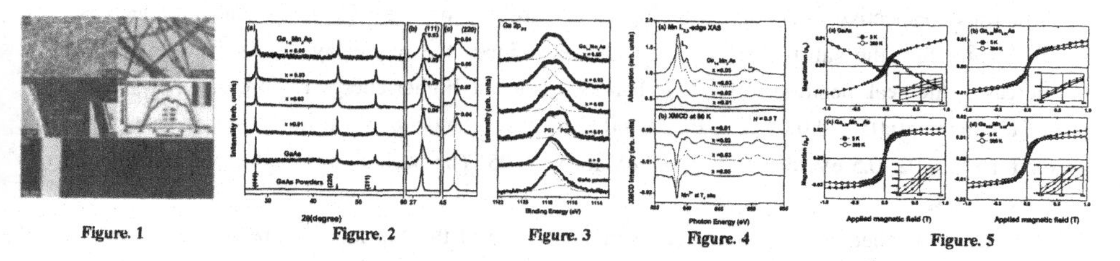

Figure. 1 Figure. 2 Figure. 3 Figure. 4 Figure. 5

Figure 1a shows the SEM micrograph of the high-density 2% Mn-doped GaAs ($Ga_{0.98}Mn_{0.02}As$) NWs grown on the substrates. The TEM image explicitly reveals their smooth surface and average diameter of 60 nm (Figure 1b). The Au catalytic-nanoparticles are frequently attached to the tips, providing evidence for the VLS growth mechanism. The lattice-resolved TEM image and corresponding FFT ED pattern of a selected nanowire reveal that it is composed of single-crystalline zinc blende structured GaAs (Figure 1c and two insets). It shows a spacing between neighboring (111) planes of ca. 3.2 Å, which is consistent with that of the bulk materials (a=5.654 ; JCPDS Card no. 80-0016). The EDX line-scanning analysis indicates the Mn content (x = [Mn]/([Ga]+[Mn])) of the individual nanowire to be about x = 0.02±0.01, with a homogeneous distribution being observed along the cross-section, as shown in Figure 1d. The inset corresponds to its high-angle annular dark field (HAADF) scanning TEM (STEM) image. The STEM image and its EDX mapping for a selected $Ga_{0.95}Mn_{0.05}As$ NW are shown in Figure 1e. The Mn element is distributed homogenously over the whole nanowire with an average Mn content of 0.05±0.01.

The high-resolution XRD patterns of the undoped GaAs NWs and $Ga_{1-x}Mn_xAs$ NWs are displayed in Figure 2a. All of the GaAs NW samples have a highly crystalline nature without the presence of other phases. Figures 2b and 2c display the magnified (111) and (220) peaks, respectively. The peak position shifts to a higher angle, most significantly when the Mn content is 0.02-0.03. The maximum shifts, $\Delta(2\theta)$, of the (111) and (220) peaks are 0.04 and 0.05 degrees, respectively. Their peak width becomes broader as the Mn concentration increases.

The feature of the fine-scanned Ga 2p peaks was strongly dependent on the Mn content. Figure 3 shows the fine-scanned Ga $2p_{3/2}$ peak. The peak of the $Ga_{1-x}Mn_xAs$ NWs is broader in width and more asymmetric, as compared to that of the undoped one. Furthermore, as the Mn content increases, the peak position shifts to the higher energy region. The peak can be deconvoluted into two bands at ~1117 (PG1) and ~1118 (PG2) eV. The peak position. The binding energy of the Ga atoms bonded to the As atoms would be expected to appear at a lower energy compared to that of the Ga atoms bonded to dangling bonds or defects (usually bonded to more electronegative O atoms). Therefore, the PG1 and PG2 bands can be assigned to the Ga-As and Ga-O bonding structures, respectively. As the Mn dopes, the area % of the PC2 band tends to decrease, suggesting that the substitution of the Ga sites reduces the oxide layers. As the Mn content increases, the peak position of the PG1 band shifts to the higher energy region; 0.2 eV x = 0.02; 0.3 eV for x = 0.03 and 0.05. The PG2 band also show a tendency to shift to a higher-energy; 0.3 eV for x = 0.02; 0.5 eV for x =0.03; 0.3 eV for x =0.05. The peak shows the highest energy shift for x = 0.03.

In order to further investigate the electronic structure of the Mn ions, we performed XAS and XMCD $(\Delta\rho = \rho^+ - \rho^-)$ measurements at the Mn $L_{2,3}$ edges. Figures 4a and 4b show the Mn $L_{2,3}$ edge XAS and XMCD spectra of the $Ga_{1-x}Mn_xAs$ (x = 0.01, 0.02, 0.03, and 0.05) NWs, respectively, measured at 80 K. The spectra, which result from the Mn 2p→3d dipole transition, are divided roughly into the L_3 ($2p_{3/2}$) and L_2 ($2p_{1/2}$) regions.

The peak features and positions of the XAS spectra suggest that the doped Mn atoms are in an ionic state, not in a metallic state such as metallic Mn clusters. The negative signal of the XMCD L_3 peak can be clearly observed, indicating the contribution of Mn^{2+} at the tetrahedral sites, substituting for the Ga ions in GaAs, which is consistent with the results obtained for Mn-doped GaAs films. These Mn^{2+} ions would be responsible for the hybridization, which results in the ferromagnetism of the GaAs NWs.

The magnetic moment (M) versus magnetic field (H) curves at 5 and 300 K were measured by the SQUID magnetometer. The nanowires were separated from the substrates. All of the samples exhibit ferromagnetic hysteresis behavior with a diamagnetic and/or paramagnetic background signal (not deducted in the figures). Figure 5a corresponds to the field-dependent M-H curve of the undoped GaAs NWs. Hysteresis occurs with a saturation field of 0.1 T at both temperatures. The saturated magnetization is estimated to be $0.005\mu_B$ at 5 K .The inset displays the curve in the vicinity of H = 0, indicating that the coercive field (H_C) is 200 Oe and the remanence (M_r) is $0.0015\mu_B$ at 5 K. Therefore, the existence of ferromagnetism at 5 and 300 K is clearly proven by the coercivity and remanence, and relatively low saturation field, suggesting that the T_C value is at least 300 K.

Figures 5b, 5c, and 5d display the M-H curves of the $Ga_{0.99}Mn_{0.01}As$, $Ga_{0.97}Mn_{0.03}As$, and $Ga_{0.95}Mn_{0.05}As$ NWs, respectively, consistently showing hysteresis at 5 and 300 K.

2. Mn doped TiO_2 nanowires

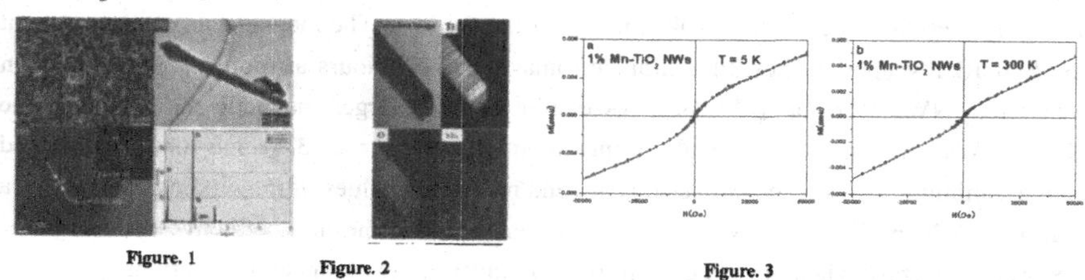

Figure. 1 Figure. 2 Figure. 3

Figure 1a shows the SEM micrograph of the high-density 1% Mn-doped TiO_2 NWs grown on the Ti foil. The TEM image explicitly reveals their surface and average diameter of 150 nm (Figure 1b). The lattice-resolved TEM image and corresponding FFT ED pattern of a selected nanowire reveal that it is composed of single-crystalline teragonal structured TiO_2 (Figure 1c and insets). It shows a spacing between neighboring (1-10) planes of ca. 3.2 Å, which is consistent with that of the bulk materials (a=4.594 ; JCPDS Card no. 86-0147). The EDX spectrum analysis indicates the Mn content (x = [Mn]/([Ti]+[Mn])) of the individual nanowire to be about x = 0.01, as shown in Figure 1d.

Figure 2 shows the EELS elemental distribution of the Ti, Mn, and O atoms over the whole nanowires, showing the homogeneous distribution of Mn.

The magnetic moment (M) versus magnetic field (H) curves at 5 and 300 K were measured by the SQUID magnetometer. The nanowires were separated from the Ti foil. $Ti_{10.99}Mn_{0.01}O$ NWs exhibit ferromagnetic hysteresis behavior with a diamagnetic and/or paramagnetic background signal (not deducted in the figures). Figure 3a and 3b correspond to the field-dependent M-H curve of the $Ti_{10.99}Mn_{0.01}O$ NWs. Hysteresis occurs with a saturation field of 0.1 T at both temperatures.

CONCLUSIONS

Single-crystalline $Ga_{1-x}Mn_xAs$ (x = 0, 0.01, 0.02, 0.03, and 0.05) NWs were grown by the vapor transport method, using the evaporation of $GaAs/MnCl_2$. They all consist of single-crystalline zinc blende GaAs crystals grown uniformly with the [111] direction, irrespective of the level of Mn doping. The EDX data reveal that the Mn dopes homogeneously over the entire nanowires. We investigated their electronic structures and magnetic properties by XRD, XPS, XAS, XMCD, and MPMS. The XRD data reveal that the lattice constant decreases with increasing MN doping level, most significantly at x = 0.03. The positions of the XPS Ga and Mn 2p peaks shift to the higher energy region when x is increased. We explained the higher-energy peak shift in terms of the hybridization between the Mn^{2+} ions and holes, which was suggested to be a crucial feature in determining the ferromagnetism. The XAS and XMCD data reveal that the Mn^{2+} ions are dominantly present at the Td sites, substituting for the Ga ions. The magnetization measurement using SQUID reveals that they all exhibit ferromagnetic behaviours at room temperature, even the undoped NWs. The $Ga_{1-x}Mn_xAs$ NWs exhibit a much larger magnetic moment than the undoped NWs, and consistently exhibit a maximum value at x = 0.03, which is well correlated with the minimum value of the lattice constant and maximum values of the binding energy of Ga and Mn XPS 2p peaks at this Mn content. We conclude that the most effective ferromagnetic nanowires are formed when the Mn content is 0.03, under our experimental conditions.

High-density TiO_2 NWs were synthesized on a large area of Ti substrates, by the thermal oxidation of Ti foil, combined with vapor transport of Mn. HRTEM images and FFT ED patterns confirm that the single-crystalline TiO_2 nanocrystals grown along the [110] direction. EDX and EELS mapping and spectrum reveal the homogeneous 1% and 3% Mn doping along the whole nanowires. Tomography measurement reveal that the nanowires are actually nanobelts with a width of ~ 500 nm. The 1% Mn-doped TiO_2 NWs consisted of pure TiO_2 phase. XAS and XMCD measurement of 1% Mn-doped TiO_2 NWs confirmed that the Mn^{2+} ions substitute into the tetrahedral (T_d) coordinated sites. Magnetization measurement revealed that 1% Mn-doped TiO_2 nanobelts exhibit room-temperature ferromagnetic behaviors.

REFERENCES

1 (a) Ohno, H. Science 1998, 281, 951. (b) Ohno, Y.; Young, D. K.; Beschoten, B.; Matsukura, F.; Ohno, H.; Awschalom, D. D. Nature (London) 1999, 402, 790. (c) Ohno, H.; Chiba, D.; Matsukura, F.; Omiya, T.; Abe, E.; Dietl, T.; Ohno, Y.; Ohtani, K. Nature (London) 2000, 408, 944.

2 Fiederling, R.; Keim, M.; Reuscher, G.; Ossau, W.; Schmidt, G.; Waag, A.; Molenkamp, L. W. Nature (London) 1999, 402, 787.
3 Wolf, S. A.; Awschalom, D. D.; Buhrman, R. A.; Daughton, J. M.; von Molnár, S.; Roukes, M. L.; Chtchelkanova, A. Y.; Treger, D. M. Science 2001, 294, 1488.
4 Ohno, H.; Shen, A.; Matsukura, F.; Oiwa, A.; Endo, A.; Katsumoto, S.; Iye, Y. Appl. Phys. Lett. 1996, 69, 363.
5 Oiwa, A.; Matsukura, F.; Endo, A.; Hirasawa, M.; Iye, Y. Ohno, H.; Katsumoto, S.; Shen, A.; Sugawara, Y. Solid State Commun. 1997, 103, 209. (x dependence)
6 Dietl, T.; Ohno, H.; Matsukura, F.; Cibert, J.; Ferrand, D. Science 2000, 287, 1019.
7 Sanvito, S.; Ordejon, P.; Hill, N. A. Phys. Rev. B 2001, 63, 165206.
8 Erwin, S.C.; Petukhov, A. G. Phys. Rev. Lett. 2002, 89, 227201.
9 Jungwirth, T.; Konig, J.; Sinova, J. Phys. Rev. B 2002, 66, 012402.
10 (a) Mahadevan, P.; Zunger, A. Phys. Rev. B 2003, 68, 075202. (b) Mahadevan, P.; Zunger, A.; Sarma, D. D. Phys. Rev. Lett. 2004, 93, 177201.
11 Bergqvist, L.; Korzhavyi, P. A.; Sanyal, B.; Mirbt, S.; Abrikosov, I, A.; Nordström, L. E.; Smirnova, A.; Mohn, P. Svedlindh, P.; Eriksson, O. Phys. Rev. B 2003, 67, 205201.
12 Krstajic, P. M.; Peeters, F. M.; Ivanov, V. A.; Fleurov, V.; Kikoin, K. Phys. Rev. B 2004, 70, 195215.
13 Schulthess, T. C.; Temmerman, W. M.; Szotek, Z.; Butler, W. H.; Stocks, G. M. Nat. Mater. 2005, 4, 838.
14 Kitchen, D.; Richardella, A.; Tang, J. M.; Flatte, M. E.; Yazdani, A. Nature 2006, 442, 436.
15 Edmonds, K. W.; Bogusławski, P.; Wang, K. Y.; Campion, R. P.; Novikov, S. N.; Farley, N. R. S.; Gallagher, B. L.; Foxon, C. T.; Sawicki, M.; Dietl, T.; Buongiorno Nardelli, M.; Bernholc, J. Phy. Rev. Lett. 2004, 92, 037201.
16 Nazumul, A. M.; Amemiya, T.; Sugahara, S.; Tanaka, M. Phy. Rev. Lett. 2005, 95, 017201-1.

Mater. Res. Soc. Symp. Proc. Vol. 1144 © 2009 Materials Research Society 1144-LL19-13

ZnO/Al_2O_3 Core-Shell Nanorod Arrays: Processing, Structural Characterization, and Luminescent Property

C. Y. Chen,[1] C. A. Lin,[1] M. J. Chen,[2] G. R. Lin,[1] and J. H. He[1,*]
[1]*Institute of Photonics and Optoelectronics, and Department of Electrical Engineering, National Taiwan University, Taipei, 10617 Taiwan*
[2]*Department of Materials Science and Engineering, National Taiwan University, Taipei, 10617 Taiwan*

ABSTRACT

We reported the aqueous chemical method to fabricate the well-aligned ZnO/Al_2O_3 core-shell nanorod arrays (NRAs). The shell is composed of α-Al_2O_3 nanocrystals in amorphous Al_2O_3 layers. The photoluminescence (PL) measurements showed that the enhancement of near-band-edge emission in ZnO NRAs arrays due to the addition of Al_2O_3 shell was observed. The Al_2O_3 shell layer resulting in flatband effect near ZnO surface leads to a stronger overlap of the wavefunctions of electrons and holes in the ZnO core, further enhancing the NBE emission.

INTRODUCTION

Due to the ultrahigh surface-to-volume ratio characteristics of nanowires/nanorods (NWs/NRs) [1], significant progress has been made for the fabrication of various electronic, optoelectronic, and sensor devices [2-6]. Since the performance of a nanodevice critically depends on the surface of nanomaterials, the investigation of surface effect is vitally important for fully utilizing the advantages offered by nanomaterials.

ZnO has been devoted to novel optoelectronic devices duo to their unique properties such as large excitation binding energy (60 meV), near ultraviolet UV emission, transparent conductivity, and piezoelectricity [7]. Engineering of the ZnO NWs/NRAs-based device through surface modification can maximize the effects or benefits provided by nanostructures. For practical applications, the growth of controlled NRAs is vitally important for applications such as field emitters [8], nanogenerator [9], solar cell [10], and nanolasers [11]. The most popular method used for growth of ZnO NRAs is to use Au as catalyst and a single crystal substrate, such as a-plane alumina. The close lattice and symmetry match between the substrate surface and the ZnO results in epitaxial vertical growth of NRAs [12]. Moreover, various complex nanostructures [13-17] have been synthesized with the modification of optical properties. However, a study on the ZnO/nanocrystal-embedded (NC-embedded) Al_2O_3 core-shell NR arrays (NRAs) has not been investigated. ZnO/Al_2O_3 core-shell structure is attractive because it is wafer-scale assembly by all aqueous chemical methods. In addition, Al_2O_3, being a high-k dielectric material, may also act as the gate dielectric of ZnO nanowire transistors [18]. Since the bandgap of Al_2O_3 is larger than that of ZnO and the refractive index of Al_2O_3 is lower than that of ZnO, it suggests that the optical property of ZnO can benefit from Al_2O_3 coating [17].

In this work, we reported the chemical method, advantageous for low reaction temperature, low cost, minimum equipment requirement, atmospheric pressure and product homogeneity, to fabricate the wafer-size ZnO/Al_2O_3 core-shell NRAs. The shells consist of α-Al_2O_3 NCs embedded in amorphous Al_2O_3 layers. The structural characterization of ZnO/Al_2O_3 NRAs was performed using field emission scanning electron microscope (SEM), and transmission electron microscope (TEM). The photoluminescence (PL) measurements were performed to determine the optical properties of ZnO/Al_2O_3 core-shell NRAs. The near-band-edge (NBE) emission in ZnO NRAs was found to be enhanced with the addition of Al_2O_3 shells due to the flatband effect near the surface of ZnO.

EXPERIMENT

The ZnO NRAs were synthesized on Si (001) substrate by a hydrothermal method. A thin film of zinc acetate was spin coated on the substrate with a solution of 5 mM zinc acetate dehydrate in ethanol. The ZnO seed layer of 5-10 nm thickness was formed after baking in air for 20 min. The ZnO NRAs were grown in 100 ml of aqueous solution, containing 10 mM zinc nitrate hexahydrate and 2.5 ml ammonia solution, at 95 °C for 2 hours. The Al_2O_3 shell layers on the ZnO NRAs were fabricated after ZnO NRAs spin-coated with 5 mM $AlCl_3$ were annealed at 220 °C for 8 hours in air ambient. In order to vary the thickness of the shell layer, ZnO NRAs were spin-coated 10 times (sample A), 15 times (sample B), and 20 times (sample C) with 5 mM $AlCl_3$, respectively.

Morphological studies of ZnO NRAs and ZnO/Al_2O_3 core-shell NRAs were performed with a JEOL JSM-6500 field emission SEM and a JEOL 3000F field emission TEM. The PL measurements were performed in air at room temperature and in a liquid helium bath cryostat in a temperature range from 14 K to 120 K using a He-Cd laser (photon energy = ~3.8eV) as an excitation source.

DISCUSSION

Figures 1(a) and 1(b) show the typical SEM images of ZnO/Al_2O_3 core-shell NRAs (sample C). The diameter and the length of the well-aligned NRAs are in the ranges of 80-100 nm and 2.5-3.0 μ m, respectively. To further analyze the structure of NRAs, TEM characterization was performed. After the growth process, the substrate-bound NRAs were mechanically scrapped, sonicated in ethanol and deposited on silica-coated copper grids for TEM characterization. A low-magnification TEM image of a nanorod from sample C as shown in figure 1(c) indicates uniformity in both of the core and shell layers. One can see that an approximately 10-nm-thick shell layer was formed for sample C. Figure 1(d) shows a high resolution TEM (HRTEM) image of the core from sample C, which confirms that the phase is wurtzite-structured ZnO. The measured interplanar distance of 0.26 nm corresponds to the ZnO (0002) planes, indicating that the NRAs grew preferentially along the *c*-axis direction (the left inset in Figure 1(d)). The right inset in Figure 1(d) shows that the shell layers are NC-embedded amorphous layer. The NCs embedded in amorphous layer are identified as α-Al_2O_3. According to the EDS analysis, the amorphous layers are composed of Al and O. In addition, TEM characterization shows that samples A and B exhibit ~5-nm-thick and ~7-nm-thick Al_2O_3 shell layer with the Al_2O_3 NCs.

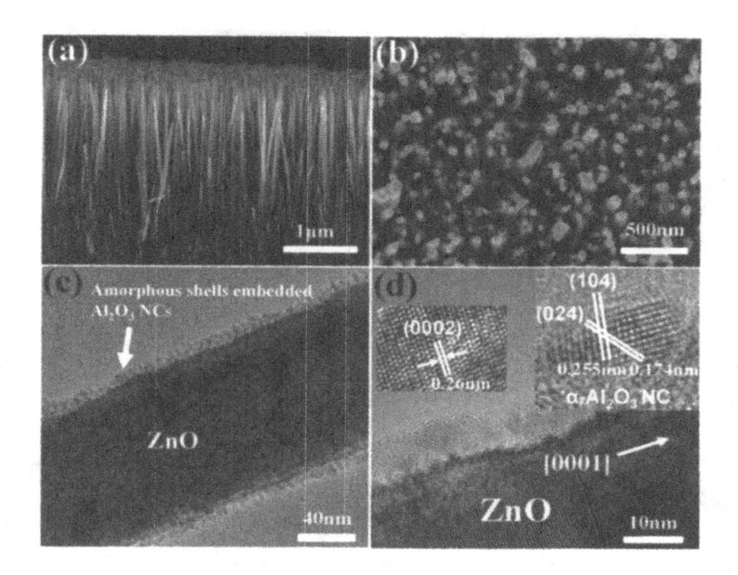

Figure 1. (a) A cross-sectional SEM image, (b) A top-view SEM image, (c) A TEM image, and (d) A HRTEM image of ZnO/Al_2O_3 core-shell NRAs (sample C). The left and right insets in (d) are the HRTEM images of ZnO core and Al_2O_3 shell, respectively.

A control sample only consisting of ZnO NRAs was prepared to measure the PL property without the influence of Al_2O_3 shell layer. Figure 2 shows typical PL spectrum of ZnO NRAs without Al_2O_3 shell layer exhibited a NBE emission at ~3.24 eV and a deep level emission at ~2.2 eV related to the defects [19]. With the presence of Al_2O_3 shell, the NBE is greatly enhanced. However, the change in the intensity of deep level emission was less pronounced for the core-shell NRAs as compared with the control sample. It was found that while the thickness of a shell layer was increased up to 10 nm, the intensities of NBE and deep level emission were qualitatively the same with the intensity for the 5-nm-thick shell layer. Furthermore, NBE peak positions were not varied with Al_2O_3 shell layer coating.

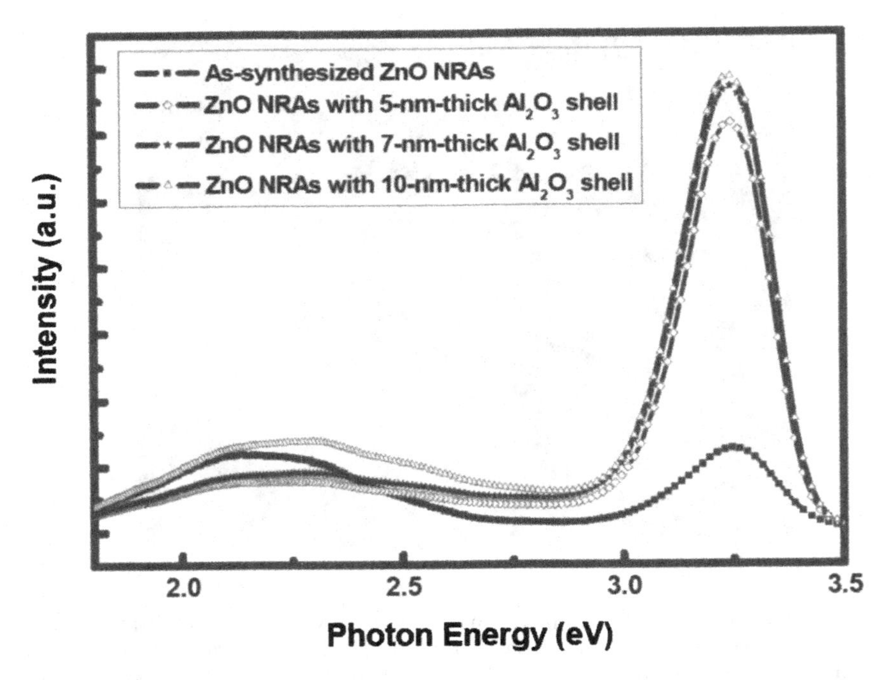

Figure 2. PL spectra of the ZnO NRAs and ZnO/Al_2O_3 core-shell NRAs at room temperature

It has been known that the recombination of deep levels responsible for visible emission in ZnO nanostructures is located at the surface (<30 nm in thickness) [20]. The nature of amorphous materials cannot suppress the surface defect emission since the shell layers did not provide significant surface passivation for the ZnO NRAs. It was found that α-Al_2O_3 NCs embedded in amorphous layer contributed little towards surface passivation. As a result, there is no significant change in the intensity of deep level emission of ZnO after amorphous Al_2O_3 layer coating. The origin of the enhanced NBE emission in core-shell NRAs can be explained by the following model. For an uncoated ZnO NRAs, oxygen molecules adsorb near the surface of ZnO NRAs, trap the free electrons from conduction band, and become the charged oxygen molecules (O_2^-, O^-, or O^{2-}) [21]. This widens the depletion layer, leading to an upward band bending around the surface [22]. This can result in a separation of photo-generated electron–hole pairs [17]. It has been reported that the thickness of depletion layer is ~20 nm for ZnO NRAs [1]. The depletion layers are expected to have a large influence on the distribution of photocarriers for ZnO NRAs with diameter smaller than 100 nm in the present study. Charged oxygen molecules near the surface are expected to be one of the oxygen sources for the formation of Al_2O_3 shell using $AlCl_3$. This causes a reduction of the surface traps and further lowers the surface band bending of ZnO cores. This results in a stronger overlap of the wavefunctions of electrons and holes in the ZnO cores, further enhancing the NBE emission of the ZnO NRAs. This model can be interpreted that the enhancement of NBE emission due to a change of the local electronic structure of a ZnO surface is insensitive to the thickness of Al_2O_3 shell.

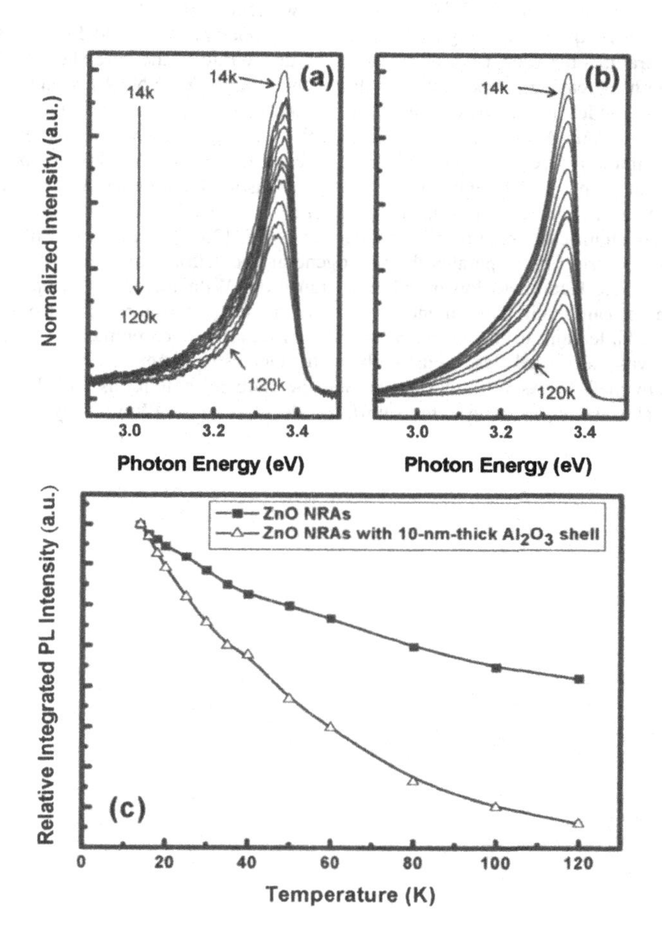

Figure 3. Temperature-dependent PL spectra of ZnO (100-nm diameter) NRAs (a) and ZnO (100-nm diameter)/Al_2O_3 (10-nm thickness) core-shell NRAs (b) measured (c) Spectrally integrated PL intensity normalized with the PL intensity at 14 K as a function of temperature for 100-nm ZnO NRAs and ZnO/Al_2O_3 core-shell NRAs.

Temperature-dependent PL measurements were further investigated to characterize the optical properties of ZnO/Al_2O_3 core-shell NRAs. An excitation density was ~22 mW cm^{-2}. Figures 3(a) and 3(b) show typical temperature-dependent PL spectra of the ZnO NRAs and the ZnO/Al_2O_3 core-shell NRAs (samples C). The PL peak position of the two samples was invariable with temperatures. ZnO NRAs with or without Al_2O_3 shell exhibit the different behaviors of thermal quenching. Figure 3(c) shows the relative integrated PL intensity of the NRAs at temperatures between 14 K and 120K. The decay rate of the core-shell ZnO NRAs in PL intensity with increasing temperature was faster than that of ZnO NRAs, which is explained by the following model. First, carriers have two channels to recombine: radiative transition and nonradiative transition. As carriers without sufficient energy cannot overcome the barrier, surrounding nonradiative center, to nonradiatively recombine at relatively low temperature [23], nonradiative transition can be suppressed, leading to carrier recombination through radiative channels. With increasing temperature, the probability of recombination through nonradiative centers becomes significant, resulting in the quench of PL [23]. For the bare ZnO NRAs, the surface built-in electric field separates the photogenerated electron–hole pairs and accumulates holes at the surface, leading to lower radiative transition. With increasing temperature, more accumulated holes can obtain thermal energy to overcome the potential resulted from the surface band banding [21], leading to increasing probability of radiative recombination. As a result, the PL intensity was decreased less sensitively with increasing temperature. In the case of ZnO/Al_2O_3 core-shell NRAs, ZnO NRs approach the flatband conditions and thus, few holes accumulation takes place, resulting in the significant decrease in the PL intensity with increasing temperature.

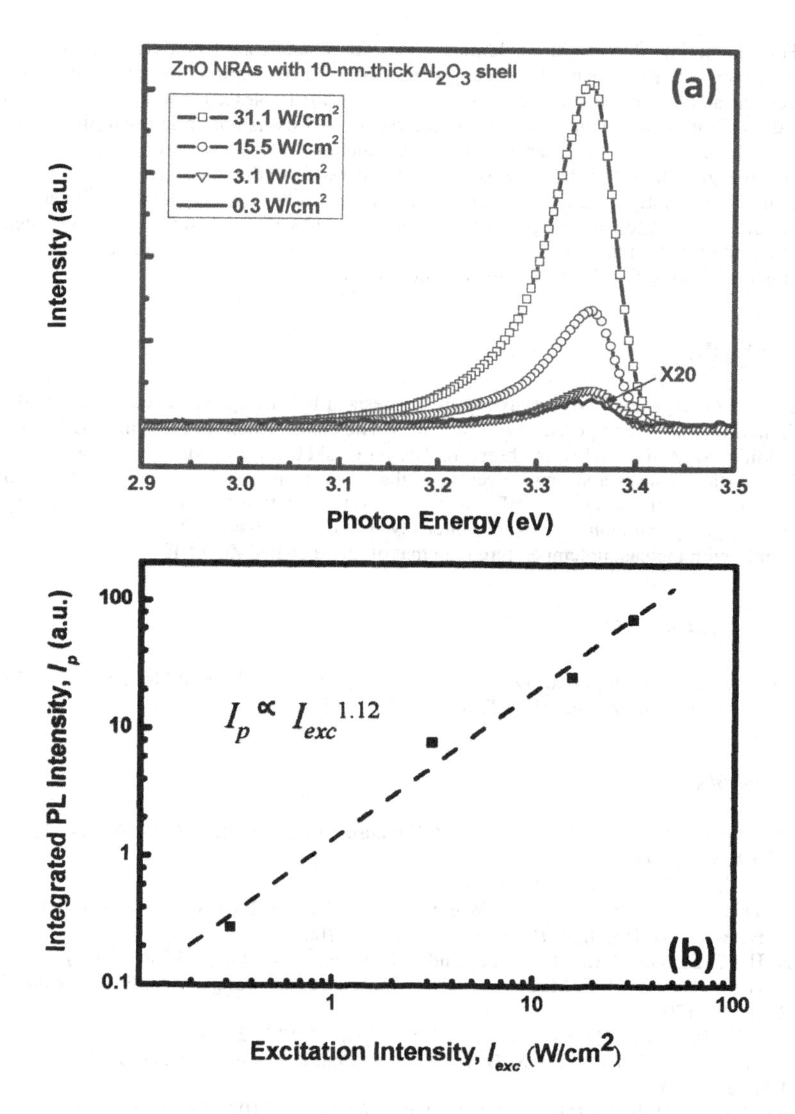

Figure 4. (a) Excitation intensity-dependent PL spectra and (b) The integrated PL intensity as a function of excitation intensity for the ZnO/Al_2O_3 core-shell NRAs (samples C)

For examining the origin of luminescence from semiconductors, excitation intensity dependence of the PL spectra has been performed. The excitation intensity-dependent PL measurement at 14K on the ZnO/Al_2O_3 core-shell NRAs (samples C) is shown in Figure 4(a). The peak position of the main UV emission located at 3.35 eV is not varied with the excitation intensity. The integrated PL intensity (I_p) as a function of the excitation intensity (I_{exc}) is indicated in Figure 4(b). The dash line is the least-squares fit to the measured data shown by the black squares. The integrated PL intensity is linearly dependent on the excitation intensity. No evidence of saturation for the integrated PL intensity was observed when the excitation intensity was increased by more than 2 orders of magnitude. This shows that the excitonic transitions are the dominant origin of the observed luminescence at 14K [24-26].

CONCLUSIONS

ZnO/Al_2O_3 core-shell NRAs have been synthesized by the aqueous chemical method. The amorphous shells were composed of α-Al_2O_3 NCs. The enhancement of NBE in ZnO NRAs due to the addition of Al_2O_3 shell was observed. The Al_2O_3 shell layer, leading to to flatband effect near ZnO surface, causes a stronger overlap of the wavefunctions of electrons and holes in the ZnO core, further enhancing the NBE emission. The temperature-dependent PL measurements confirmed this speculation since PL intensity of the core-shell NRAs decreases more significantly with increasing temperature than that of the uncoated ZnO NRAs.

ACKNOWLEDGMENTS

The research was supported by the National Science Council Grant No. NSC 96-2112-M-002-038-MY3 and NSC 96-2622-M-002-002-CC3.

REFERENCES

1. D. H. Weber, A. Beyer, B. Völkel, A. Gölzhäuser, E. Schlenker, Bakin A. and A. Waag, *Appl. Phys. Lett.* **91** 253126 (2007)
2. D. Appell, *Nature (London)* **419**, 553 (2002)
3. X. F. Duan, Y. Huang, Y. Cui, J. F. Wang and C. M. Lieber, 2001 *Nature (London)* **409**, 66
4. J. H. He and C. H. Ho, *Appl. Phys. Lett.* **91** 233105 (2007)
5. J. H. He, C. L. Hsin, J. Liu, L. J. Chen and Z. L. Wang, *Adv. Mater.* **19** 781 (2007)
6. J. H. He, Y. H. Lin, M. E. McConney, V. V. Tsukruk, Z. L. Wang and G. Bao, *J. Appl. Phys.* 102, 084303 (2007)
7. G. Yi, C. Wang and W. I. Park, *Semicond. Sci. Technol.* **20** S22 (2005)
8. K. F. Huo , Y. M. Hu, J. J. Fu, X. B. Wang, P. K. Chu, Z. Hu and Y. Chen, *J. Phys. Chem.* C. **111** 5876 (2007)
9. X. D. Wang, J. H. Song, J. Liu and Z. L. Wang, *Science* **316** 102 (2007)
10. M. Law, L. E. Greene, J. C. Johnson, R. Saykally and P. D. Yang, *Nat. Mater.* **4** 455 (2005)
11. M. H. Huang, S. Mao, H. Feick, H. Q. Yan, Y. Y. Wu, H. Kind, E. Weber, R. Russo and P. D. Yang, *Science* **292** 1897 (2001)

12. J. H. He, J. H. Hsu, C. W. Wang, H. N. Lin, L. J. Chen and Z. L. Wang, *J. Phys. Chem.* B **110** 50 (2006)
13. C. Hsu, Y. Lin, S. Chang, T. Lin, S. Tsai and I. Chen, *Chem. Phys. Lett.* **411** 221 (2005)
14. S. Z. Li, C. L. Gan, H. Cai, C. L. Yuan, J. Guo, P. S. Lee and J. Ma, *Appl. Phys. Lett.* **90** 263106 (2007)
15. W. I. Park, J. Yoo, D. W. Kim, G. C. Yi and M. Kim, *J. Phys. Chem.* B **110** 1516 (2006)
16. Y. H. Park, Y. H. Shin, S. J. Noh, Y. Kim, S. S. Lee, C. G. Kim, K. S. An and C. Y. Park, *Appl. Phys. Lett.* **91** 012102 (2007)
17. J. P. Richters, T. Voss, D. S. Kim, R. Scholz and M. Zacharias, *Nanotechnology* **19** 305202 (2008)
18. X. H. Zhang, B. Domercq, X. D. Wang, S. Yoo, T. Kondo, Z. L. Wang and B. Kippelen, *Org. Electron.* **8** 718 (2007)
19. A. B. Djurisic and Y. H. Leung, *Small* **2** 944 (2006)
20. I. Shalish, H. Temkin and V. Narayanamurti, *Phys. Rev.* B **69** 245401 (2004)
21. G. H. Schoenmakers, D. Vanmaekelbergh, and J, J, Kelly, *J. Phys. Chem.* **100** 3215 (1996)
22. Y. J. Lin and C. L. Tsai, 2006 *J. Appl. Phys.* **100** 113721
23. J. I. Pankove, *Optical processes in semiconductors* (Prentice-Hall) p.166 (1971)
24. J. E. Fouquet and A. E. Siegman, *Appl. Phys. Lett.* **46** 280 (1985)
25. Z. C. Feng, A. Mascarenhas and W. J. Choyke, *J. Lumin.* **35** 329 (1986)
26. C. L. Yang, J. N. Wang, W. K. Ge, L. Guo, S. H. Yang and D. Z. Shen, *J. Appl. Phys.* **90** 4489 (2001)

Mater. Res. Soc. Symp. Proc. Vol. 1144 © 2009 Materials Research Society 1144-LL19-16

Structural and Optical Properties of Pseudobinary Wurtzite Alloy Nanowires

S. Joon Kwon[*,1,2] , Young-Jin Choi[2], Kyoung-Jin Choi[2], Dong-Wan Kim[2], and Jae-Gwan Park[2]
[1]*Department of Chemical Engineering, Massachusetts Institute of Technology, MA 02139*
[2]*Nano Science and Technology Division, Korea Institute of Science and Technology (KIST), Seoul, Korea*

ABSTRACT

Pseudobinary CdS_xSe_{1-x} and ZnS_xSe_{1-x} ($0 \leq x \leq 1$) alloy nanowires were synthesized on an Au-coated Si substrate by pulsed laser deposition (PLD) process. The synthesized alloy nanowires were single crystalline hexagonal wurtzite structures. Both the lattice constant and the unit cell volume of each alloy nanowire were linearly correlated with the composition, x, and therefore satisfying the Vegard's law. Bandgap of each alloy nanowire measured by photoluminescence (PL) also changed linearly with the composition implying its tunability in the spectral region over a range of 1.75-2.45 eV (CdS_xSe_{1-x}) and 2.66-3.50 eV (ZnS_xSe_{1-x}) with more systematic controllability. This linear scaling behavior of the bandgaps of each alloy nanowire was clearly distinguished from the case of thin films made of the same materials and some case of the ZnS_xSe_{1-x} alloy nanowires reported by another group mainly due to the inner strain relaxation in the confined one-dimensional structure of the nanowires.

INTRODUCTION

Among II-VI compound materials, CdS/CdSe and ZnS/ZnSe have received intensive research attention due to their possible application to the fabrication of optoelectronic device. To achieve tunability in the wide spectral range of UV-visible light, a solid solution comprising CdS/Se and ZnS/Se can provide an appropriate scheme ranging from 1.75-2.45 and 2.66-3.50 eV. In the present study, we report on the structural and optical properties of wurtzite crystalline CdS_xSe_{1-x} and ZnS_xSe_{1-x} ($0 \leq x \leq 1$) alloy nanowires synthesized by PLD. The lattice constants of the alloy nanowires are linearly correlated with the composition. Bandgaps of each nanowire are strongly dependent on the composition, in contrast to the thin films. We discuss the uniquenss of the optical properties of the alloy nanowires which implies the bandgap tunability for optoelectronic devices.

EXPERIMENTAL

CdS_xSe_{1-x} and ZnS_xSe_{1-x} alloy nanowires were synthesized on Au (2 nm-thick)-coated Si (100) substrates by PLD. To prepare the target materials, mixtures of CdS/CdSe and ZnS/ZnSe powders (purity > 99.9%, Aldrich) were pressed using cold pressing methods to form CdS_xSe_{1-x} and ZnS_xSe_{1-x} pellet with the designed mole fraction such that $x = 0.00, 0.25, 0.50, 0.75, 1.00$ and then sintered at 500°C for 2h. It should be noted that the real mole fractions of CdS/CdSe and ZnS/ZnSe in the synthesized nanowires were slightly different from the designed mole fractions. KrF excimer laser (Lambda Physik, Compex-205, KrF radiation at $\lambda = 248$ nm) was ablated on the stoichiometric targets placed at the centre of a quart tube. Well-developed nanowires were obtained at the down-stream side of the chamber after 20 min of ablation at 700 to 800°C. Scanning electron microscopy (SEM) (FEI NOVA, 15 kV) was used to examine the morphology of the resulting structures and x-ray diffraction (XRD) (Bruker D8, Cu Ka radiation with λ = 1.5046Å) and transmission electron microscopy (TEM) (FEI Tecnai F20, 200 kV) to measure the crystallinity. The compositions of the alloy nanowires were measured by energy-dispersive analysis of X-ray (EDAX)-equipped TEM. Photoluminescence (PL) was excited at room temperature by a large-frame, nitrogen ion laser to measure the bandgap energy of the synthesized alloy structure.

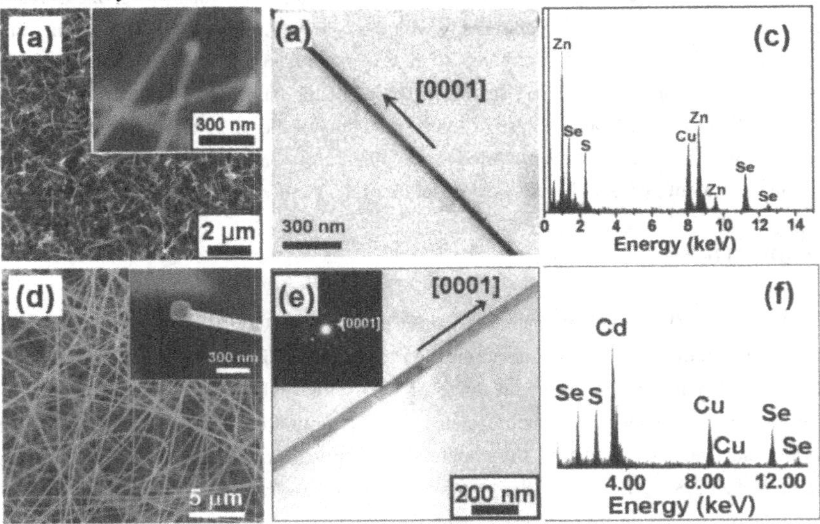

Figure 1. The 1st low for CdS_xSe_{1-x} and the 2nd low for ZnS_xSe_{1-x} alloy nanowires. From left SEM and expanded FESEM images (a,d), TEM and HR-TEM images with selected area electron diffraction (SAED) pattern (b,e), and EDAX analysis of the alloy nanowires (c,f).

RESULTS and DISCUSSION

Fig. 1 shows typical images of the CdS_xSe_{1-x} (Fig. 1(a),(b),(c)) and ZnS_xSe_{1-x} (Figure 1.(d),(e),(f)) alloy nanowires grown at 700-800°C with x = 0.500 and 0.475, respectively. The nanowires had diameters ranging from 50 to 200 nm and lengths of 100 μm. The magnified images in the insets of the Fig. 1(a) and (d) show the alloy tip of Au/CdS_xSe_{1-x} and Au/ZnS_xSe_{1-x} at the end of the nanowires indicating the vapor-liquid-solid (VLS) mechanism governed the nanowire growth. As shown in Fig. 1(b) and (e), the growth axis of the CdS_xSe_{1-x} and ZnS_xSe_{1-x} nanowires was parallel to the [0001] direction and the nanowires were single crystalline.

The XRD patterns provided in Fig. 2 show the crystallinity of the different CdS_xSe_{1-x} and ZnS_xSe_{1-x} alloy nanowires. The series of diffraction patterns indicated that the alloy nanowires were commonly hexagonal wurtzite crystalline structures without second phases. Simple calculation confirmed that the lattice constants of the *a*- and *c*-axes of the alloy nanowires scaled linearly with the x value. These scaling behaviors of the lattice parameters indicated that the CdS_xSe_{1-x} and ZnS_xSe_{1-x} alloy nanowires were complete solid solutions without structural defects satisfying Vegard's law [1].

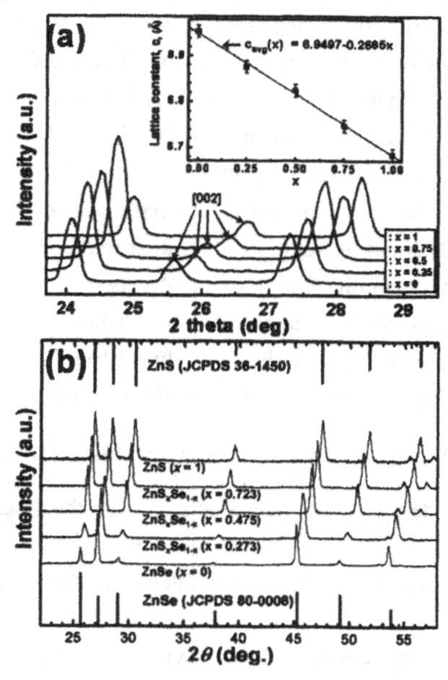

Figure 2. XRD Patterns of the CdS_xSe_{1-x} (a) and ZnS_xSe_{1-x} alloy nanowires (b). Linearly scaling behavior of the lattice constant a of the CdS_xSe_{1-x} alloy nanowires was shown in the inset of (a).

An interesting finding in the present study was the linear relation between the bandgap energy and the x value. The PL spectra of the CdS_xSe_{1-x} and ZnS_xSe_{1-x} alloy nanowires with different x values excited at room temperature are shown in Fig. 3(a) and (c). The luminescence due to excitonic emissions was measured in the CdS_xSe_{1-x} and ZnS_xSe_{1-x} alloy nanowires and the near band emission (NBE) peaks were shifted towards a large bandgap energy with increasing x value. Fig. 3(b) and (d) present the strong linear relationship between x and the bandgaps, indicating that the bowing effects of the bandgap energy variation in the alloy nanowires were nearly zero. More interestingly, the bowing effect was also reported in the cases of ZnS_xSe_{1-x} alloy nanowires grown by thermal evaporation with bowing parameter of 0.60 eV [2] and ZnS_xSe_{1-x} alloy tetrapod nanorods grown by CVD with bowing parameter of 0.64 eV.[3] In the case of thin films, as is the case of bulk materials, inner strain produced during the duration of synthesis over the plane direction of the films was not completely relaxed, therefore, bowing effect governed by this remained strain in the films should be accompanied. It is well known that this bowing effect can be described with quadratic function accompanied by bowing parameter between 0 to 0.65 eV.[4] In the event of ZnS_xSe_{1-x} alloy nanowires grown by thermal evaporation, we can suggest several reasons for the bowing effect. Firstly, inner strain might had not been completely relaxed due to composition-dependent the growth direction of the alloy nanowires, and this can result in a slight variation in the PL spectra of each nanowire. Secondly, the growth temperature of ZnS_xSe_{1-x} alloy nanowires using thermal evaporation (1100°C) was considerably higher than that employed in our experiment (700-800°C), which can give rise to additional thermal strain in the alloy nanowires resulting in a variation in the PL spectra of each nanowire. More concentrated study on the thermal effect is required. In the case of ZnS_xSe_{1-x} alloy tetrapod nanorods grown by CVD,[2] Xu *et al.* suggested that their tetrapod alloy nanorods also exhibited strong bowing effect with quadratic fitting of their experimental data. They calculated the bowing parameter, and it seemed to be in the range of typical bowing parameter of ZnS_xSe_{1-x} alloy.[3] As shown in the inset of Fig. 3(d), however, linear fitting is better for their experimental data than quadratic fitting, since the correlation coefficient calculated from the linear fitting curve ($R = 0.981 \pm 0.0898$) is larger than that from the quadratic fitting ($R = 0.966 \pm 0.1059$). Additionally, the standard deviation of the value of R for the linear fitting (0.0898) is also smaller than that for the quadratic fitting (0.1059). This result indicates that the bandgap of ZnS_xSe_{1-x} alloy tetrapod nanorods is linearly correlated with the x value, and no bowing effect subsequently.

The disappearance of the bowing effect observed in our cases is believed to be mainly due to the reduction of the bowing parameter, which is strongly dependent on the inner strain of the alloy structure. While complete relaxation of this inner strain is unnecessary for the formation of the alloy thin film, it is required to form a stable, uniform, grown in a uniaxial direction, single crystalline, one-dimensional alloy structure such as nanowire. In our case, the CdS_xSe_{1-x} and ZnS_xSe_{1-x} alloy nanowires had not only a single-crystalline structure but also a uniaxial elongated structure (or mono growth direction of [0001]) with a uniform diameter along the longitudinal direction without structural defects. Complete relaxation of inner strain is a prerequisite for these uniformities in the crystallinity and geometry of the alloy structure. The bowing parameter approaches zero as the short-range order parameter governed by the inner strain approaches zero. In the hexagonal wurtzite crystalline structure, complete relaxation of the inner strain reduces the short-range parameter to zero, therefore, there is no bowing effect in the CdS_xSe_{1-x} and ZnS_xSe_{1-x} alloy nanowires since the short-range order parameter governs the quadratic term in the quadratic relationship between the bandgap and the composition of the alloy structure. More profound explanation between the bowing effect and strain relaxation in the alloy nanowires can be found in our previous report.[5] The linearly correlated behavior of the bandgap observed in this study can alloy us to conclude that the synthesized CdS_xSe_{1-x} and ZnS_xSe_{1-x} alloy nanowires have more precisely predictable tunabilily. Therefore we believe that the experimentally confirmed controlling capability of the direct bandgap energy of the CdS_xSe_{1-x} and ZnS_xSe_{1-x} alloy nanowires in the spectra range covering 1.75-2.45 eV and 2.66-3.50 eV can shed some light on the possible applications of optoelectronic devices based on II-VI compound semiconductors with more practical materials design.

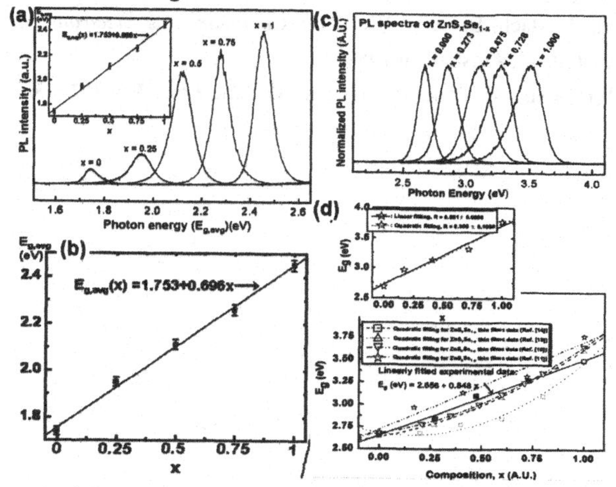

Figure 3. PL spectra of the CdS_xSe_{1-x} (a) and ZnS_xSe_{1-x} alloy nanowires (c) measured at room temperature. Note that bandgaps calculated from the PL spectra of the different CdS_xSe_{1-x} (b) and ZnSxSe1-x alloy nanowires (d) scale linearly with x.

CONCLUSIONS

We studied the structural and optical properties of single-crystalline wurtzite CdS_xSe_{1-x} and ZnS_xSe_{1-x} nanowires synthesized by PLD. The alloy nanowires grown satisfying the VLS mechanism, and were comprised of a complete solid solution of CdS/CdSe and ZnS/ZnSe. The variation in the lattice constants of the CdS_xSe_{1-x} and ZnS_xSe_{1-x} alloy nanowires due to the increased value of the composition showed a satisfactory agreement with Vegard's law. The bandgap energy of the CdS_xSe_{1-x} and ZnS_xSe_{1-x} alloy nanowires measured by room-temperature PL spectra also showed linear correlations with the composition in the ranges of 1.75-2.45 eV and 2.66-3.50 eV, respectively. The disappearance of the bowing effect observed in both cases was due to the complete relaxation of the inner strain in the single-crystalline one-dimensional structure with uniaxial and uniform geometry. The optical properties of the CdS_xSe_{1-x} and ZnS_xSe_{1-x} alloy nanowires promised ease tunability of bandgap emission and also showed the strong chance of application in optoelectronic devices based on II-VI compound semiconductors.

REFERENCES

1. L. Vegard, *Z. Phys.* **1921**, *5*, 17.
2. M. Wang, G.T. Fei, Y.G. Zhang, M.G. Kong, L.D. Zhang, *Adv. Mater.* **2007**, *19*, 4491.
3. H. Xu, Y. Liangm Z. Liu, X. Zhang, S. Hark, *Adv. Mater.* **2008** (in-press).
4. H. Hartmann, R. Mach, B. Selle, *Wide Gap Compounds as Electronic Materials (Ed. E. Kaldis)*, North-Holland, Amsterdam **1982**.
5. S.J. Kwon, Y.J. Choi, J .H. Park, I.S. Hwang, J.G. Park, *Phys. Rev. B* **2005**, *72*, 205312.

Mater. Res. Soc. Symp. Proc. Vol. 1144 © 2009 Materials Research Society 1144-LL19-21

Novel Inorganic DC Lateral Thin Film Electroluminescent Devices Composed of ZnO Nanorods and ZnS Phosphor

Tomomasa Satoh, Yuki Matsuzawa, Hiroaki Koishikawa, and Takashi Hirate
Faculty of Engineering, Kanagawa University, 3-27-1 Rokkakubashi, Kanagawa-ku, Yokohama, Japan

ABSTRACT

A novel inorganic thin-film electroluminescence (TFEL) device exhibiting bright EL emission when driven by a low DC voltage is demonstrated. The DC-TFEL device is based on a composite layer in which aluminum-doped ZnO nanorods are vertically embedded in ZnS:Mn as an EL phosphor. The DC driving voltage is then applied laterally to the composite layer via two side electrodes set 3.5 mm apart. The aluminum-doped ZnO nanorods were synthesized on a glass substrate by low-pressure thermal chemical vapor deposition combined with laser ablation, and the composite layer was formed by electron-beam deposition of ZnS:Mn onto the ZnO nanorods. The thickness of the composite layer was about 160 nm. After electrical modification to breakdown a basal conduction ZnO path, the lateral DC-TFEL device exhibited bright EL emission without avalanche breakdown, achieving a luminance of 747 cd/m^2 at 4200 V with a luminous efficiency of 9.2×10^{-3} lm/W.

INTRODUCTION

The commercial market for flat panel displays is presently dominated by liquid crystal and plasma technology following considerable progress in performance and cost reduction for these technologies. Organic light-emitting diodes (OLEDs) have recently attracted much attention as a promising technology for the next generation of flat panel displays. In contrast, the development of inorganic thin-film electroluminescent (TFEL) devices, also a promising display technology, lacks momentum. The slow progress in TFEL devices can primarily be attributed to the lack of an efficient blue EL phosphor, as well as the high AC voltage required to drive the devices. As efficient inorganic blue EL phosphors such as $BaAl_2S_4$:Eu have now been developed [1,2], only the problems of driving voltage remain to be addressed. A lower DC voltage is highly desirable for driving TFEL devices. The requirement for a high AC voltage stems from the basic operation of inorganic EL phosphors, which require high electric fields of the order of 10^6 V/cm to induce emission. However, such high electric fields cause avalanche breakdown of the EL host material. Conventional inorganic TFEL devices therefore have a thin-film stack structure consisting of a phosphor layer sandwiched between two insulators, which necessitates an AC driving voltage. To obtain EL emission at lower voltages, the phosphor layer and the insulator layers must be made thinner, and the driving voltage must be applied vertically to the stack via one opaque and one transparent electrode. However, a driving voltage of higher than 200 V is required even for the minimum practical thicknesses of the phosphor and insulator layers. It is thus difficult to realize further decreases in the electric field required to induce emission of the EL active layer. To achieve a low DC driving voltage, it is therefore necessary to develop a new TFEL device structure. To the best of our knowledge, however, no TFEL devices driven by a low DC voltage have been reported in the literature.

In previous work, we achieved a reduction in the driving voltage of AC-TFEL devices by developing a new structure involving the insertion of a ZnO nanorod layer in the conventional AC-TFEL structure [3]. The geometry of ZnO nanorods adjacent to an EL phosphor layer generate local electric field enhancement at the apices of the nanorods by the concentration of electric fluxes. This field enhancement lowers the required driving voltage at intermediate luminance and below. The effect of the ZnO nanorods on the driving voltage was found to be very sensitive to the properties of ZnO nanorods [3], and doping of the ZnO nanorods with aluminum was found to be effective in augmenting this effect [4]. However, the driving voltage remains AC, and the insertion of the ZnO nanorod layer results in a decrease in maximum luminance, attributable to the small volume in which the field enhancement occurs.

In the present study, another novel TFEL device is developed utilizing the local field enhancement effect generated by the geometry of conductive nanorods. The new device includes a composite layer in which conductive nanorods are vertically embedded into an EL phosphor material, as shown Figure 1(a). Contrary to conventional AC-TFEL devices, the driving voltage for the proposed device is applied laterally to the composite layer via two side electrodes, making it possible to achieve EL emission by applying a lower DC voltage. The high electric field (ca. 2×10^6 V/cm) required to induce emission by the EL phosphor materials is obtained in a local volume adjacent to the side surface of the nanorods, as shown Figure 1(b). Stable DC operation is expected to be achieved because the collision time of electrons with the host lattice on the short-current path between nanorods is reduced, preventing avalanche breakdown. This TFEL device is the first such device to be stably driven using a DC voltage.

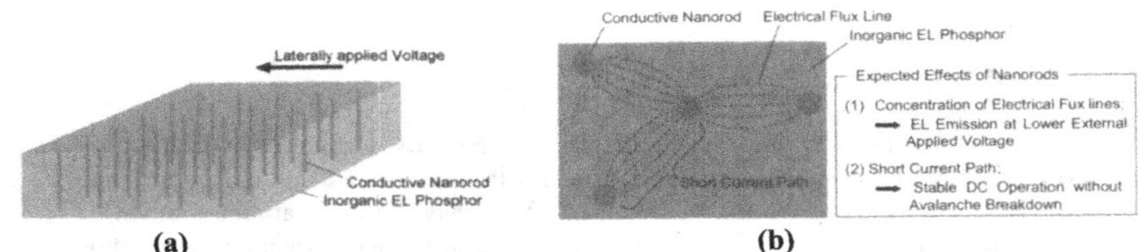

Figure 1. (a) Ideal structure of composite layer. (b) Top view of composite layer and expected effect of conductive nanorods

EXPERIMENT

A schematic diagram of the lateral DC-TFEL device structure developed in this study is shown in Figure 2. Aluminum-doped ZnO nanorods were synthesized on a 26×37 mm glass substrate by low-pressure thermal chemical vapor deposition (CVD) method combined with laser ablation using metal Zn vapor and O_2. Aluminum doping was performed by the laser ablation of Al_2O_3 during CVD growth of the ZnO nanorods. The details of the growth method have been reported previously [5,6]. The ZnS:Mn EL phosphor material was then deposited by electron-beam deposition such that the ZnS:Mn infiltrated into the spaces between ZnO nanorods (Fig 2(a)). A laterally continuous composite layer comprised of ZnS:Mn and ZnO nanorods was thus formed at the upper end of the nanorods. The substrate was finally diced into 3.0×3.5 mm chips, and electrodes were applied to opposing cross-sectional surfaces using an ultrasonic soldering system (USM-IV and CERASOLZER#246, Kuroda Techno) (Fig. 2(b)). The electrical and luminescent characteristics of the DC-TFEL device were measured under application of a DC voltage at a pressure of 5×10^{-6} torr.

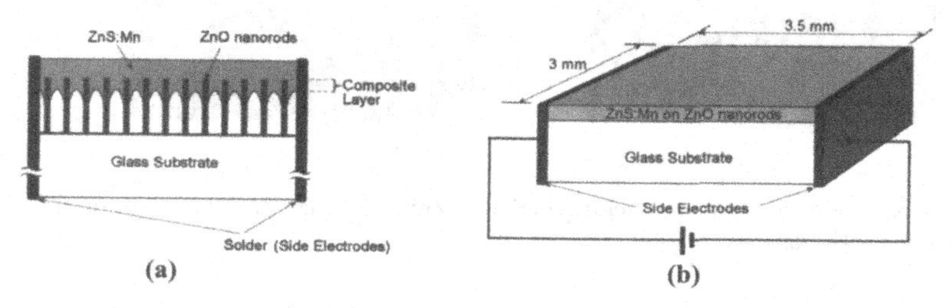

Figure 2. Schematic diagram of lateral DC-TFEL structure

RESULTS AND DISCUSSION

Figure 3 shows scanning electron microscopy (SEM) images of aluminum-doped ZnO nanorods prepared for the lateral DC-TFEL device. The ZnO nanorods were flat-tipped and were hexagonal prisms in shape, with diameter of 50–80 nm, length of ca. 1.1 μm, and density of approximately 13 rods/μm^2 on the substrate. The aluminum concentration in the nanorods was below the detection limit of energy-dispersive X-ray (EDX) analysis (< 0.3 at%). The verticality of the nanorods with respect to the substrate was relatively poor. An un-preferred continuous layer of ZnO, which is formed before the growth of the ZnO nanorods in the present growth process, existed under the ZnO nanorods. The nanorods were therefore not fully electrically isolated from one another.

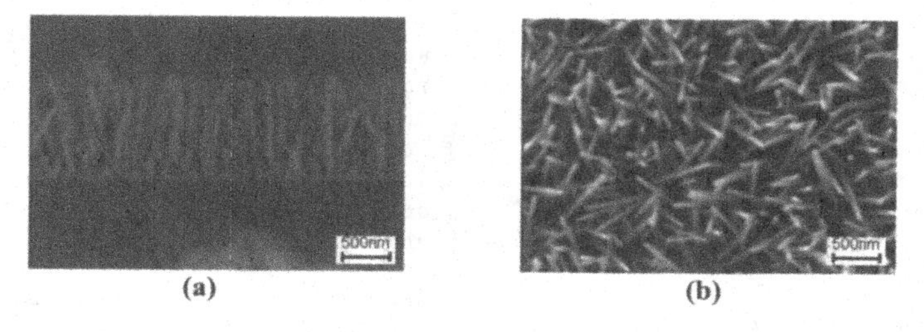

Figure 3. SEM images of Al-doped ZnO nanorods prepared for lateral DC-TFEL device. (a) Cross-sectional view, (b) top view.

Figure 4 shows SEM images of the device after deposition of ZnS:Mn on the ZnO nanorods. The laterally continuous ZnS:Mn layer at the upper end of the nonorods can be clearly seen. Considering the length of the ZnO nanorods, the thickness of the upper composite layer is estimated to be approximately 160 nm. The manganese concentration in the ZnS:Mn was determined by EDX to be approximately 0.6 at%.

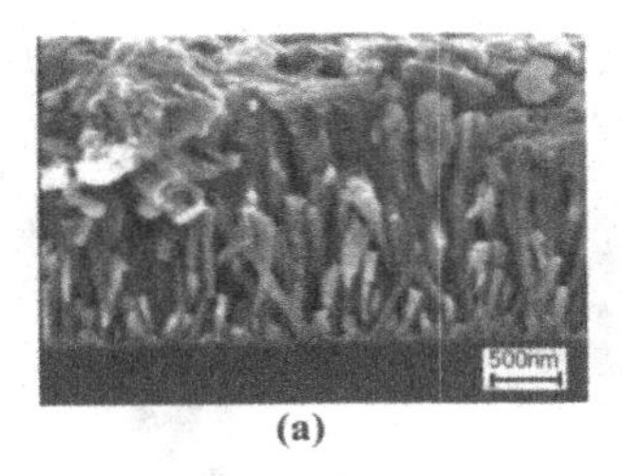

(a)

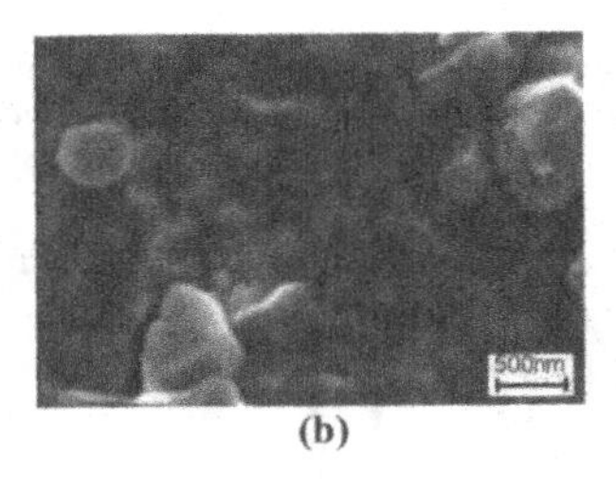

(b)

Figure 4. SEM images of device after deposition of ZnS:Mn on the ZnO nanorods. (a) Cross-sectional view, (b) top view.

The current–voltage (I–V) characteristics of the ZnO nanorod layer and the as-prepared lateral DC-TFEL device using a comparable ZnO nanorod layer under a laterally applied DC voltage are shown in Figure 5. These characteristics were measured under a gradually increasing voltage (5 V/step and 0.5 s/step). The I–V curve for the ZnO nanorod sample was very linear, corresponding to a resistance of 1.04 MΩ. This result shows that the lateral electrical isolation of ZnO nanorods was poor, attributed to the continuous ZnO layer formed at the base of the nanorods. The as-prepared lateral DC-TFEL device also displays a linear I–V curve up to 700 V, and the resistance in this voltage region was similar to that of ZnO nanorod sample. Above 100 V, the I–V curve for the lateral DC-TFEL became unstable, and the current droped abruptly at 1405 V from 7.0 mA to 0.16 mA. During these measurements, no EL emission was observed. With further increases in voltage up to 5000 V, the unstable I–V characteristic continued, and no EL emission was observed. The power consumption at the end of the linear range and immediately before the large change in current was 0.72 W and 9.8 W, respectively.

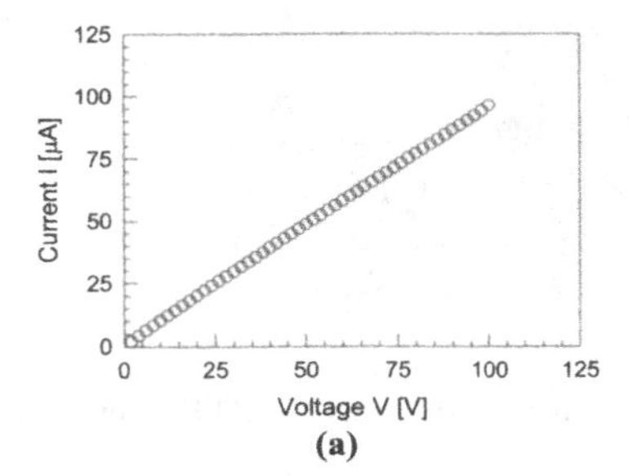

(a)

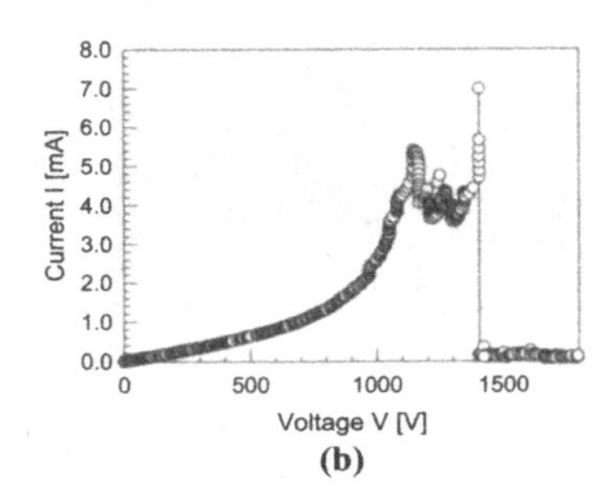

(b)

Figure 5. Current-voltage characteristics for (a) ZnO nanorods sample, and (b) as-prepared lateral DCEL device

The instability of the I–V curve is considered to be caused by local thermal destruction of the continuous ZnO layer, the ZnO nanorods, and the ZnS:Mn in the composite layer, preventing EL emission. The high power consumptions can be also be attributed to the low resistance in the continuous ZnO layer. It was therefore deemed necessary to eliminate the lateral conduction path of the continuous ZnO layer in order to obtain EL emission.

To do so without severely damaging the composite layer, an electrical formation prpcess was made to the as-prepared lateral DC-TFEL devices. The electrical formation prpcess is that a 2 s pulse of high voltage (2.5 kV) with a current limit of 3 mA is applied three times to the as-prepared lateral TFEL device, resting for 20 min between each application. The current, luminance (L), and luminous efficiency (η) characteristics of the electrically modified lateral DC-TFEL devices are shown in Figure 6. The current now increases linearly with voltage up to approximately 180 V, corresponding to a resistance of 906 MΩ, three orders of magnitude greater than before electrical treatment. At higher voltages, the current curve increased non-linearly with some instability, accompanied by uniform EL emission by the ZnS:Mn. The threshold voltage, determined as the voltage at which the luminance reaches 1 cd/m^2, is 750 V. The luminance increased with some instability with increasing voltage, reaching a luminance of 747 cd/m^2 and luminous efficiency of 9.2×10^{-3} lm/W at 4200 V. It is noted that this voltage is much lower than that required to induce emission in the lateral DC-TFEL device without ZnO nanorods, which is estimated to be 750 kV based on the spacing between electrodes and the electric field strength (2×10^6 V/cm) required to induce ZnS:Mn emission. Although the luminous efficiency of the present device is three orders of magnitude lower than that of a conventional AC-TFEL device having a ZnS:Mn phosphor layer, the realization of high luminance without critical breakdown using a DC driving voltage represents an important development.

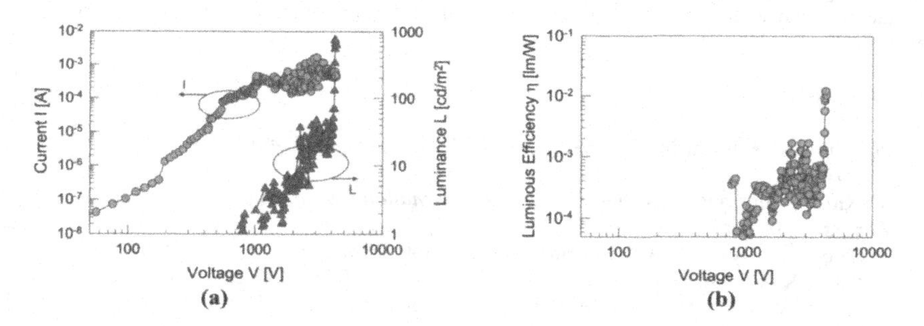

Figure 6. Electrical and EL emission characteristics of the lateral DC-TFEL device after electrical modification. (a) I–V and L–V characteristics (b) η–V characteristic.

The electrical formation process is considered to partially break the current path through the continuous ZnO layer by increasing the resistance of the layer. However, it is also expected that this electrical process caused appreciable damage to the composite layer. The instabilities of the I–V and L–V curves above the threshold voltage are probably attributable to both progression of the destruction of the current path through the continuous ZnO layer, and progression of damage in the composite layer. The achievement of lower driving voltage of the present devices than that of devices without ZnO nanorods is probably attributable to the effect of local field enhancement adjacent to the side surface of the ZnO nanorods. Aluminum doping level of ZnO nanorods in the present devices is low (< 0.3 at%). We consider that even so low aluminum doping augments the effect of the field enhancement, because equivalent aluminum doping level have augmented the effect for another AC-TFEL devices using ZnO nanorods reported in Reference [4]. Effects of aluminum concentration of ZnO nanorods on performance of the lateral DC-TFEL devices are our future work. Carrying out it, however, is difficult because the power of the laser-ablation for aluminum doping considerably affects on growth morphology of ZnO such as density of ZnO nanorods and thickness of continuous layer at the base of ZnO nanorods.

CONCLUSIONS

A new inorganic DC-TFEL device was proposed based on a composite layer in which conductive aluminum-doped ZnO nanorods are vertically embedded in a ZnS:Mn EL phosphor material. The DC driving voltage is applied laterally to the composite layer at an electrode spacing of 3.5 mm. The devices exhibit bright EL emission without critical avalanche breakdown under DC voltage operation, achieving luminance of 747 cd/m^2 at 4200 V. This operating voltage is two orders of magnitude lower than that required for EL devices without the ZnO nanorods. The present DC-TFEL device has considerable promise as a new type of inorganic EL device with simplified voltage requirements (DC and stable low voltage). This is the first successful demonstration of a lateral DC-TFEL device, and further improvements in lowering the driving voltage and increasing efficiency are expected to be made by optimizing the morphology and properties of the ZnO nanorods. In particular, the spacing between electrodes should be reduced, and the laterally conductive path at the base of the ZnO nanorods should be eliminated.

REFERENCES

1. N. Miura, M. Kawanishi, H. Matsumoto, and R.Nakano, *Jpn. J. Appl. Phys.*, **38**, L1291 (1999)
2. Y. Xin, T. Hunt, and J. Acchione, *2004 SID International Symposium Digest of Technical Papers*, 1138 (2004)
3. T. Satoh, H. Miyashita, A. Nishiyama, and T. Hirate, *Proc. ASID'06*, 512 (2006)
4. T. Satoh, K. Takizawa, T. Shinao, and T. Hirate, *Proc. EURODISPLAY 2007*, 364 (2007)
5. T. Hirate, S. Sasaki, W. Li, H. Miyashita, T. Kimpara, and T. Satoh, *Thin Solid Films* **487**, 35 (2005)
6. H. Miyashita, T. Satoh and T. Hirate, *Superlattices and Microstructures* **39**, 67 (2006)

Mater. Res. Soc. Symp. Proc. Vol. 1144 © 2009 Materials Research Society 1144-LL20-11

Directed assembly of nanowires using silicon grooves and localized surface treatments

Sabrina Habtoun, Christian Bergaud, Monique Dilhan, David Bourrier
LAAS-CNRS, University of Toulouse, 7, avenue du Colonel Roche, 31077 Toulouse Cedex 4, France

ABSTRACT

A collective method to localize, align and address electrically unidimensional nanostructures by using only photolithographic steps was developed. It relies on one hand on the structuration of a silicon substrate with anisotropic etching, resulting in V-shaped grooves leading to both alignment and localization by their geometry. On the other hand, specific surface treatments (inside and outside the grooves) are carried out in order to increase the alignment yield. The influence of the groove geometry and of the nature of the surface treatments was studied. With these optimized protocols, gold nanowires were aligned and addressed electrically, the characterization of single nanowires showing an ohmic behaviour.

INTRODUCTION

The properties of nanostructures and the need to address them electrically into micro/nanosystems have made their directed assembly a field of extensive research. The alignment of 1D nanostructures can be induced by electrical[1], magnetic[2], capillary forces[3]. Our work focuses on the directed assembly of bottom-up fabricated nanowires on patterned silicon substrates. The patterning consists of anisotropic etching of V-grooves and variation of the contact angle[4] inside and outside the grooves. In this approach, the capillary assembly induced by the variation of the contact angle of the substrate is associated with the geometric effect of the trenches, which act as microchannels. Moreover, the geometry of the anisotropic trenches allows a good control of the position of the aligned nanowires (at the bottom) without having to resort to e-beam lithography. This method is a collective one, in which the nanowires are addressed statistically. Therefore, maximizing the yield of the alignment is a significant issue to make this assembly efficient.

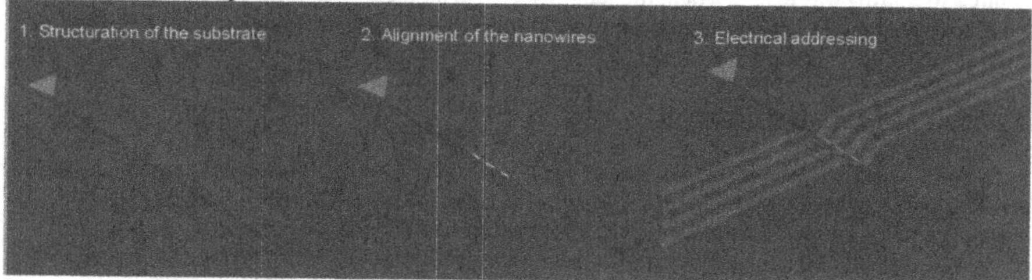

Figure 1 - The v-groove-based method to align nanowires

EXPERIMENT

Fabrication of the nanowires

Gold nanowires were fabricated by electrodeposition inside commercially available porous alumina membranes[5] coated with copper. They have a diameter of 250 nm and a length of 5 µm. They were then dispersed in water after dissolution of the template in sodium hydroxyde.

Fabrication of the template

All the steps described here were processed by photolithography. A 4'' silicon wafer (p-type, <100>) was coated with silicon nitride (LPCVD) in which an array of micron-wide lines were etched (RIE). The silicon nitride acted as a mask for the anisotropic etching of silicon (using TMAH), thus creating V-trenches. After a wet etching of the silicon nitride, the silicon substrates were treated with various SAMs, with liquid-phase surface treatments, in order to tune the contact angle between the nanowire solution and the silicon, thanks to photolithography. The aim was to deposit a hydrophobic SAM outside the trenches (OTS, octadecyltrichlorosilane) to increase the yield of the alignment, and a hydrophilic one inside them (APTMS, aminopropyltrimethoxysilane) to improve the ordering of the nanowires. The patterning of the OTS can be done using photoresist, whereas the APTMS can only be patterned afterwards, with the OTS acting as a mask.

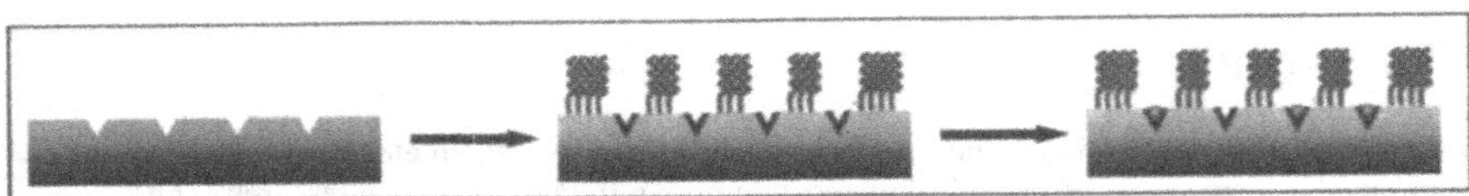

Figure 2 - Example of substrate patterning, with a hydrophobic SAM outside the grooves and a hydrophilic one inside

In order to investigate the influence of the contact angle tuning, the following treatments were applied to the substrates:

- a reference sample without any surface treatment (SiO2 sample)
- samples with a hydrophobic SAM outside the trenches (OTS) and either no treatment inside (SiO2) or a hydrophilic SAM inside (APTMS)
- samples with the hydrophilic SAM everywhere (APTMS)

Moreover, the influence of the groove size was studied: first, the width of the grooves (from 2 to 20µm) was tested. Then, the length was also studied, with long grooves (2mm) designed for quantitative observations and short ones (250µm) to favor subsequent electrical addressing.

Assembly and electrical addressing

A few microliters of the solution of nanowires were deposited on the patterned substrate and left to evaporate. After the deposition and a cleaning of the SAM by O2 plasma, gold electrodes (a titanium adhesion layer of 25nm, and 300nm of gold) were deposited by lift-off above the aligned nanowires. The exposure time of the photolithography had to be adapted to the depth of the trenches. The lift-off step has to be done very carefully in order not to take the nanowires away. The addressing is statistical, therefore the yield of the alignment increases the probability of having addressed nanowires, which have to be located with a microscope prior to characterization. Then, the nanowires were characterized electrically with a probe-testing station.

DISCUSSION

Assembly of the nanowires

The reference sample (patterned silicon without further treatment) showed some nanowires localized and aligned by the geometric effect. However, the yield was very low (around 1%).

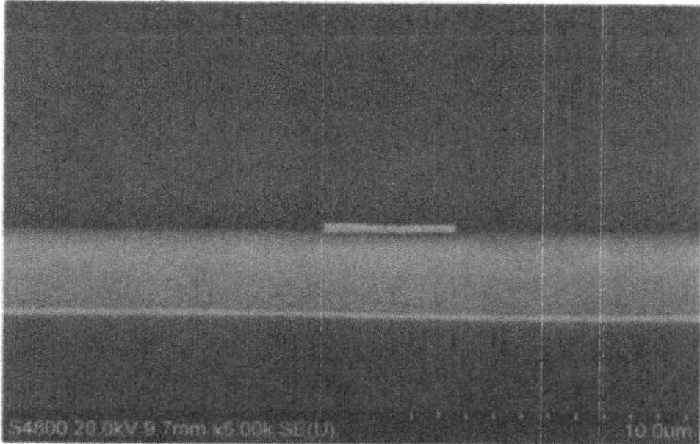

Figure 3 - Nanowire aligned at the bottom of a tilted groove

As expected, the samples with the hydrophobic SAM outside showed a great increase of the yield, which was greater than 50%. However, the quality of the alignment was quite bad, as there was a great aggregation of the nanowires even though the concentration of the solutions used was low. Moreover, many nanowires stuck to the side of the grooves instead of the bottom, probably because of the difference in contact angle which was too important. The presence of an hydrophilic treatment inside the grooves made little difference in this case.

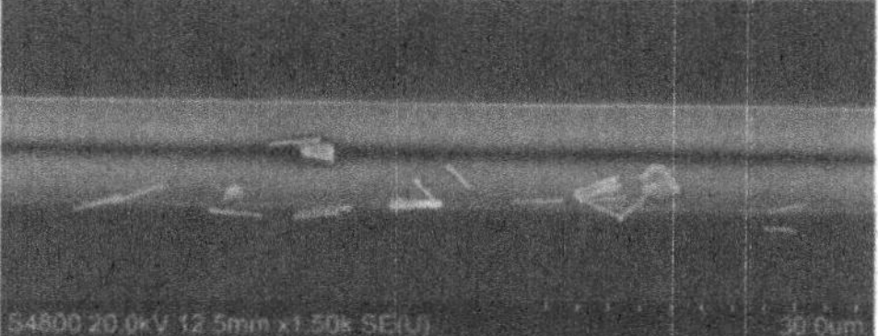

Figure 4 - Nanowires inside a groove with OTS outside, APTMS inside

The best results were obtained when the hydrophilic SAM was put everywhere in the sample. The yield is lower than with OTS (around 10%), but the nanowires were well spread. This can be explained by two causes. First, the sample is less hydrophilic (contact angle ranging from 30 to 60°) than simple silicon dioxide (0 to 30°). As a consequence, the capillary flow is slower and tends to give a better organization. Then, APTMS has a strong electrostatic interaction with colloidal gold, which leads to the formation of a monolayer.

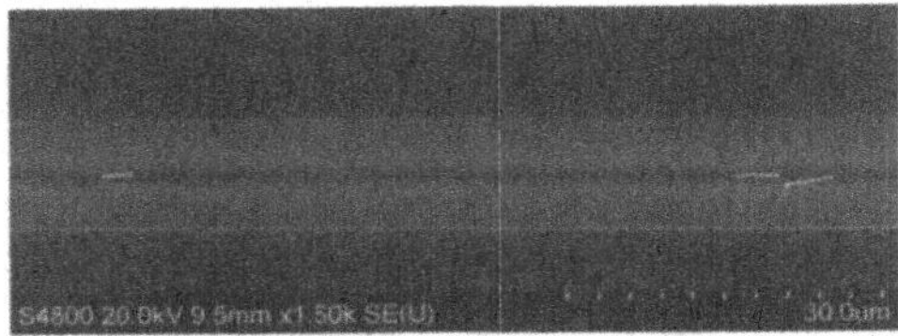

Figure 5 - Nanowires aligned inside a groove with APTMS inside and outside the patterns

The width of the patterns showed little significance in the results (actually, the yield increased with the size, but this is mainly due to the difference of area which increases the probability of a wire being near a pattern). The length of the lines did not interfere either, as it was possible to align nanowires in narrower patterns. This increases the probability to address them afterwards.

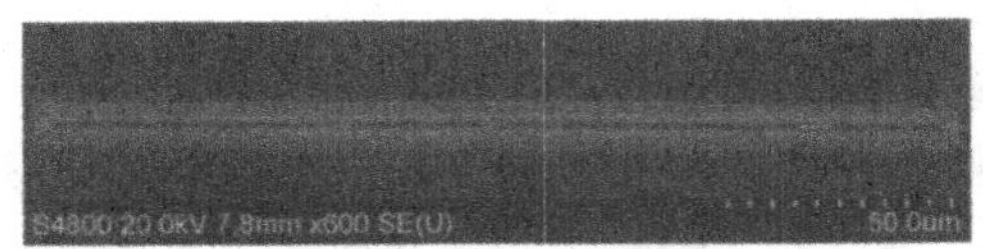

Figure 6 - Nanowires aligned in a 250μm pattern

Addressing of the nanowires

It was possible to address nanowires electrically by a lift-off of gold microelectrodes. Electrical characterization showed an ohmic behaviour, which was very reproducible.

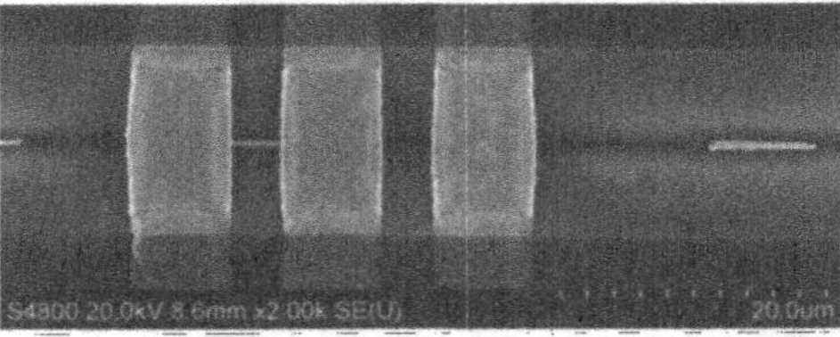

Figure 7 - Nanowires addressed electrically with photolithography-patterned electrodes

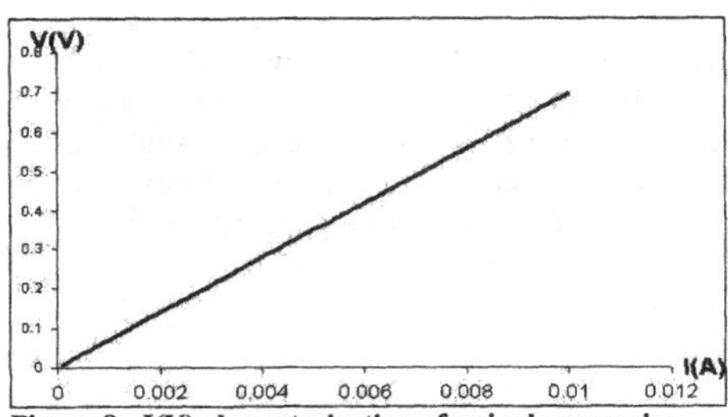

Figure 8 - I(V) characterization of a single nanowire

If the lift-off is done before the deposition of the nanowires, the gold electrodes interfere too much with the flow of the nanowire solution, and alignment is impossible. However, the deposition above the nanowires avoids the issue of contact resistance. Annealing also decreases the resistance observed (around 50Ω for the electrodes and 4Ω for the nanowires).

CONCLUSIONS

This method allowed us to align and characterize single nanowires. It is easy and low-cost, yet statistic, therefore the results depend largely on the yield of the localization. The yield can be highly increased, but at the expense of the quality of the nanowires spreading.

An improvement of the method could be a step using a photolithography resist in order to "lift-off" the nanowires which are not located in the grooves. This would allow an increase of the number of nanowires aligned in the trenches without increasing the ones between the patterns, the resist being hydrophobic enough to replace the OTS layer.

REFERENCES

1. P. A. Smith and C. D. Nordquist and T. N. Jackson and T. S. Mayer and B. R. Martin and J. Mbindyo and T. E. Malouk, Appl. Phys. Lett. **77** (9), 1399 (2000)
2. C. M. Hangarter and N. V. Myung, Chem. Mater. **17**,1320 (2005)
3. Y. Yin and Y. Lu and B. Gates and Y., J. Am. Chem. Soc. **123**, 8718 (2001)
4. S. Liu and J. Tok and J. Locklin and Z. Bao, Small **2** (12), 1448 (2006)
5. J. C. Hulteen and C.R. Martin, J. Mater. Chem. **7** (7), 1075 (1997)

Mater. Res. Soc. Symp. Proc. Vol. 1144 © 2009 Materials Research Society 1144-LL20-17

Silver Nanowires: Synthesis, Characterization and Optical Properties

Yu. A. Barnakov, H. Li, G. Zhu, M. Mayy, E. Robinson, C. Bonner and M. A. Noginov
Center for Materials Research, Norfolk State University, Norfolk, VA 23504, USA

ABSTRACT

We describe electrochemical synthesis of a bulk metamaterial, consisting of silver nanowires in Porous Anodic Alumina membrane and its characterization. We have found that the quality of the synthesized metamaterial depends on the metal used as a working electrode at the optimal conditions of the electroplating process. The dissolution of the thin layer of working electrode is occurred during the electrochemical reaction. It suggests an important role of interfacial phenomena taking place between the working electrode and the sample holder.

INTRODUCTION

In the past few years, artificial sub-wavelength metal nanostructures embedded into dielectric media, known as Metamaterials (MMs), have received enormous attention of scientific community [1,2]. They do not exist in nature and their extraordinary properties are determined by the interplay of their underlying materials properties and their structural arrangement. This opens up the possibilities to engineer materials by the tailoring and controlling of the values of relative electric permittivity, ε and magnetic permeability, μ. The interaction of electromagnetic waves with such materials results in a number of unprecedented phenomena, such as the negative refraction [3], sub-wavelength imaging (superlens and hyperlens) [4, 5], and optical cloaking [6]. Current progress towards the fabrication of MMs has been enabled by advances in nanofabrication technology. Several structures that have been proposed have been synthesized for operation in the both microwave and optical frequencies ranges, such as split ring resonators [7], paired metal stripes [8], fishnet [9]. However, until now, the creation of 3-D bulk MMs operating in the optical frequencies range has remained a major challenge because of significant optical losses and difficulties in fabrication.

Recently, it has been shown that a metamaterial based on the regular array of silver nanowires embedded into Porous Anodic Alumina (PAA) membrane can be considered as the 3D bulk photonic metamaterial [3, 10]. In this work, we report on the synthesis of this type of metamaterial.

Since it was first reported, the use of PAA membrane as template for the synthesis of nanowires has a long history [11, 12]. A range of methods for the synthesis of nanowires of different metals, including silver, gold and copper were developed [13-15]. Also in those reports, the relationship between synthesis, structure, and basic optical properties was established. A typical synthesis procedure [13] involves the deposition of a metal thin film on one of the sides of membrane, which serves as working electrode, followed by electrochemical deposition into the pores of the membrane where the pores are used to restrict the dimensions of the metal rods that form. The technique is reliable, simple and economical. However, the two main drawbacks for these methods are associated with the necessity to use either (1) thick metal film as working electrode (>100 nm), which results in non-transparency of the sample or (2) thin metal film, which, in turn, leads to an extremely low content of metal nanowires. Both makes impossible to study negative refraction in such samples.

In the present article, we describe the method of synthesis of Ag nanowires in PAA with relatively equivalent faces of the membrane and high metal filling factor.

EXPERIMENTAL DETAILS

Porous Anodic Alumina membranes (PAA) with the dimensions 1cm x 1cm x 51μm have been purchased from Synkera Technologies, Inc. Silver nanorods were synthesized in PAA via electrochemical plating. A thin gold or silver film, (~50 nm) deposited on the membrane's surface via thermal vapor deposition technique, served as the working electrode and the graphite rod played the role of the counter electrode. PAA with the side of depositing metal film was attached to the copper plate as a holder. The rest of the holder was covered with insulating tape. The mixture of aqueous solutions of Silver nitrate (1.76 M) and Boric acid (0.7 M) were used as the electrolyte. A few drops of concentric Nitric acid were added to adjust pH=2-3. The DC voltage of 2 V was applied during ~200 min deposition. Structural characterization was done by X-Ray Diffraction using Rigaku X-ray diffractometer. The compositional cross-section profile was measured by EDX spectroscopy with the e-beam spot size about 1 μm. Scanning Electron Microscope (SEM) JEOL -2050 operated at 30kV was used to observe array of silver nanowires in PAA channels. For SEM measurements, the cross-section side of the sample was etched in 1 M NaOH for 5 minutes to remove alumina walls and open up channels with silver. To facilitate the sample conductivity and prevent "charging up", thin layer of Au was deposited on the etched cross-section side of the sample. Transmission Electron Microscope (TEM) operated at 200 kV was used to image single nanowires. For TEM measurements, the sample was completely dissolved in 1M NaOH, washed and rinsed in ethanol and distillated water with following re-dispersion of black precipitation in ethanol. Few drops of this suspension were placed on the copper-carbon mesh and dried to make the sample ready for TEM observation.

SYNTHESIS

The synthesis includes two stages: (1) deposition of conductive metal on the one of the sides of membrane to form the working electrode and (2) electroplating in the standard electrochemical cell, containing electrolyte with the silver ions and boric acid. Gold, silver, and aluminum can be used as metal to form the working electrode. In the present work, in preparation of Sample A and Sample B, we use the gold and silver, correspondingly.

CHARACTERIZATION

Figure 1 shows an XRD pattern of Ag-PAA nanocomposite.

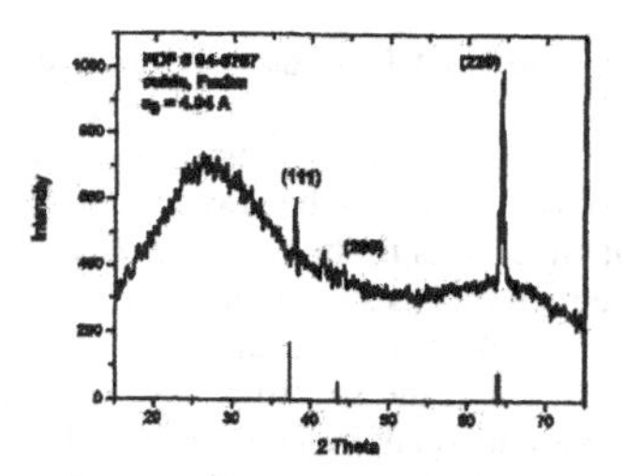

Figure 1. XRD pattern of Ag-PAA Sample A.

The broad peaks at 25 and 65 degrees correspond to the glass sample holder. The sharp intense peaks were identified as reflections (111), (200) and (220), characteristic to the cubic structure of syn - silver with Fm3m symmetry and unit cell parameter, a = 4.04 A.

As the main criteria of equivalency of the sample's faces, we used a surface roughness, R, which was estimated from Atomic Force Microscope measurements. We found that the roughness, *R*, of the Sample A and pure membrane doesn't differ dramatically. For the Sample A, the obtained values were R_s = 4.925 nm and R_e = 11.830 nm, on the solution and electrode sides, correspondingly. For the pure membrane, roughness on both sides was slightly low, R_s = 3.815 nm and R_e, = 7.253 nm. The small difference between values for R_e in pure and synthesized samples can be attributed to the presence of the small quantity of Au electrode in the form of isolating, spherical nanoparticles.

Figure 2a shows a SEM picture of the Sample A produced with the use of Au as working electrode.

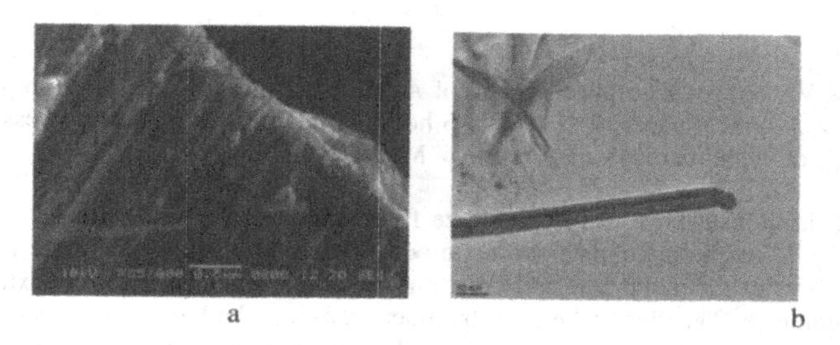

Figure 2. SEM picture of the cross-section of the Sample A (a) and TEM image of corresponding silver nanowire - gold dot at the end of it (b).

In Figure 2 two distinct shapes can clearly be seen: Ag nanowires and spherical dots, identified by EDX as Au nanoparticles. To observe the structural configuration and relationship of nanowires and dots in the respect to each other, we conducted TEM measurements of the dissolved Sample A in concentric NaOH. Figure 2b shows clearly that Au dot is attached to the Ag nanowire.

To investigate the mechanism of the formation of such structure, we performed the reference experiment, where Au deposited on the alumina membrane was exposed to the same electrochemical process in a cell but without silver ions in the solution. The optical absorption spectra of Au deposited membrane were recorded as shown in Figure 3a.

The intense absorption band centered at 550 nm, which is attributed to the surface plasmon resonance of gold nanoparticles, is clearly seen in the spectrum. We performed Mie calculation of absorption spectrum of 35 nm gold nanoparticles embedded into dielectric media with refractive index, n = 1.6 and compare it with the experimental curve. It is clearly seen in Figure 3b, that the peak positions are well matched. This is manifested the confinement of gold nanoparticles in pores of PAA. The result also correlates with TEM observation and EDX analysis. It has to be mentioned, that an intensity of the Plasmon peak reduces in absorption spectrum of the sample, which was exposed to the electrochemical treatment, as it is seen in Figure 3a (traces #3). In our opinion, this is associated with the partial removing of gold nanoparticles from membrane's surface.

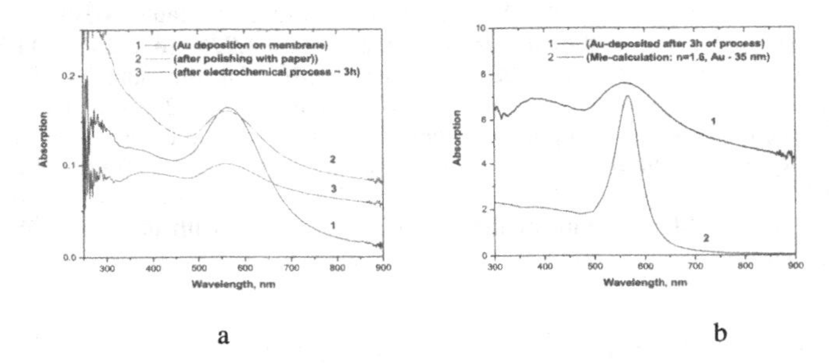

Figure 3. UV-Vis optical absorption spectra of Au-deposited membrane (a) as prepared (1), polished with optical paper (2) and after 3 hours of the electrochemical process (3); (b) comparison of experiment (1) with results of Mie calculation (2).

Based on these results we can hypothesize following conclusion. Thermal vacuum deposition of gold on the membrane surface produced two types of clusters: one which forms in the pores of the membrane and other one which loosely adsorbs at the external surface of membrane. The electrochemical treatment removes the loosely adsorbed gold clusters and but did not effect the clusters located within the pores. An electrochemical stability of these clusters is explained by their physical confinement in membrane's pores. Once, they form within the pores, they serve as nucleation centers for the electrochemical growth of the silver rods.

We attempted to eliminate the presence of Au dots by replacing of gold working electrode by a silver film. Figure 4a shows SEM picture of the cross-section of the sample electrochemically prepared with a Ag working electrode (Sample B) and TEM image of the resulting single silver nanowire (Figure 4b).

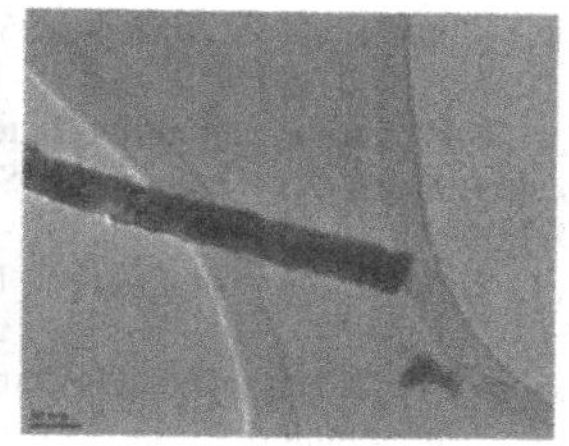

a b

Figure 4. SEM picture of the cross-section of Sample B (a) and TEM image of corresponding single silver nanowire (b).

It can be seen in Figure 4a that silver is absent within approximately 1 μm from electrode side of membrane. This is also confirmed in the EDX cross-sectional profile.

We explain this observation by processes occurring at the interface between layer of working electrode and copper plate in acidic conditions where cathodic dissolution of copper and the participation of Cu^{2+} in red-ox processes resulting in the oxidation of metal nanoparticles of working electrode and dissolution of the working electrode and the rods in the pores closest to the working electrode. More work requires elucidate this phenomenon.

CONCLUSIONS

We have demonstrated that choice of metal as working electrode significantly affects the silver filling profile in the membrane channels. The interfacial phenomena between layer of working electrode and sample's holder play a significant role towards the obtaining of the equivalent sides of metamaterial.

ACKNOWLEDGEMENTS

The work was supported by the NSF PREM grant # DMR 0611430, NASA URC grant # NCC3-1035. The authors thank Olga Trofimova for assistance with AFM and EDX measurements.

REFERENCES

1. J. B. Pendry, Phys. Rev. Lett. **85**, 3966 (2000).
2. V. Veselago, L. Braginsky, V. Shklover, C. Hafner, J. of Comput. and Theor. Nanoscience, **3**, 1, (2006).
3. J. Yao, Z. Liu, Y. Liu, Y. Wang, C. Sun, G. Bartal, A. M. Stacy, X. Zhang, Science, **321**, 930 (2008).
4. J. B. Pendry, Phys. Rev. Lett., **85**, 3966 (2000).
5. Y. Xiong, C. Sun, Z. Liu, H. Lee, X. Zhang, Science, **315**, 1686 (2007).

6. W. Cai, U. K. Chettiar, A. V. Kildishev, V. M. Shalaev, Nature Photonics, **1**, 224 (2007).
7. R. A. Shelby, D. R. Smith, S. Schultz, Science, **292**, 77 (2001).
8. V. A. Podolskiy, A. K. Sarychev, V. M. Shalaev, J. Nonlinear. Opt. Phys. Mater., **11**, 65 (2002).
9. G. Dolling, C. Enkrich, M. Wegener, C. M. Soukoulis, S. Linden, Opt. Lett., **31**, 1800 (2006).
10. M. A. Noginov, Yu. A. Barnakov, G. Zhu, T. Tumkur, H. Li, E. Narimanov, Bulk photonic metamaterial with hyperbolic dispersion, http://archiv.org/abs/0809.1028, (2008).
11. L. Piraux, J. M. George, J. F. Depress, C. Leroy, E. Ferain, R. Legras, K. Ounadjela, A. Fert, Appl. Phys. Lett., **65**, 2484 (1994).
12. D. Routkevich, T. Bigioni, M. Moskovits and J. M. Xu, J. Phys. Chem. **100**, 14037 (1996).
13. S. Bhahacharrya, S. K. Saha and D. Chakravorty, Appl. Phys. Lett., **76**, 3896 (2000).
14. G. L. Hornyak, C. J. Patrissi and C. R. Martin, J. Phys. Chem. B, **101**, 1548 (1997).
15. A. Blondel, J. P. Meier, B. Doudin and J. P. Ansermet, Appl. Phys. Lett., **65**, 3019 (1994).

Mater. Res. Soc. Symp. Proc. Vol. 1144 © 2009 Materials Research Society 1144-LL21-04

The Strength of Gold Nanowires and Nanoporous Gold

Rui Dou and Brian Derby
Materials Science Centre, The University of Manchester, Grosvenor Street, Manchester, M1 7HS, UK

ABSTRACT

We have measured the yield strength of gold nanowires with diameters in the range from 30 to 70 nm fabricated by electro-deposition into porous alumina templates. All nanowire sizes showed yield strengths much greater than polycrystalline gold with the 30 nm specimens having a yield strength of 1.4 GPa. We found no significant work hardening at plastic strains up to 30%. The strength of the nanowires as a function of wire diameter follows the same trend as has been found for the compression strength of larger gold pillars reported in the literature. TEM observations of deformed wires are consistent with mechanisms of dislocation induced deformation. The strength of nanoporous gold nanowires measured by uniaxial compression test is also reported here. Although two different mechanisms are thought to operate in gold nanowires and nanoporous gold respectively, their strengths show very similar dependence on wire or ligament diameter. However the nanoporous material shows significant strain hardening.

INTRODUCTION

The yield strength of sub-micron gold pillars measured by uniaxial compression test has been reported in many studies [1-3]. There is a significant size effect with smaller diameter pillars showing larger values of yield stress. Possible explanations for this size effect in the deformation of small crystalline structures have been reviewed by Nix *et al* and can be ascribed to two mechanisms [4], either (1) the presence of strain gradients, in which a geometrically stored population of dislocations leads to strengthening or (2) dislocation starvation, where the presence of a free surface and the associated image forces remove mobile dislocations from the structure, hence, new dislocations must be generated to maintain deformation. Because there is believed to be no strain gradients generated during the uniaxial compression test, the observed size effect in the deformation of sub-micron gold pillars has been generally ascribed to the dislocation starvation mechanism. A similar strength dependence on ligament size is also found in nanoporous gold, although in this case there are significant strain gradients during deformation.

In most reports to date, sub-micron, single crystal gold pillars are fabricated by focused ion beam (FIB) machining of bulk gold specimens with the smallest pillar diameter of 180 nm [3]. Nanoporous gold with ligament diameters in the range from 5nm to several hundred nanometers can be fabricated by selective dissolution of the Ag component from Au-Ag alloys. Because of the presence of strain gradients in the deformation of ligaments in nanoporous gold, there is a need to investigate the deformation of nanowires with diameters smaller than FIB machined gold pillars. Here we use porous alumina templates to fabricate gold nanowires, with diameters in the range 30 – 70 nm, by electro-deposition. The mechanical properties of gold nanowires were investigated by uniaxial compression testing.

EXPERIMENT

Fabrication of gold nanowire forests

Our gold nanowires were fabricated by electrodepositing gold into the pores of a highly ordered anodic aluminium oxide (AAO) template with controlled pore diameter and spacing. We used the two-step anodization process pioneered by Masuda to make hexagonally ordered porous alumina membranes [5,6]. The details of preparation of AAO templates and electro-deposition of gold nanowires have been published in Scripta Materila [7].

Scanning electron microscope (SEM) images of a representative gold nanowire forest is shown in figure 1a. The nanowires are straight and parallel with identical diameter, and height determined by the anodized template. Using transmission electron microscope (TEM), we found that the individual gold nanowires were single crystals, each with a different crystal orientation (figure 1b).

a) 

b) 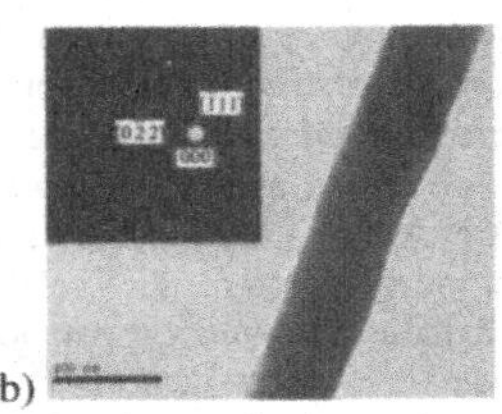

Figure 1. a) SEM image of a free standing gold nanowire forest, the inset shows a top view of the gold nanowires. b) TEM image of a 70 nm diameter single crystal gold nanowire, the inset shows a diffraction pattern of (100) projection.

Uniaxial compression test

Uniaxial compression tests on gold nanowire forests were conducted using a MTS Nanoindenter XP (MTS Nano Instruments, Oak Ridge, TN, USA), fitted with a 10 μm diameter cylindrical diamond flat punch tip. In each experiment the indenter was loaded at a constant loading rate (1 μNs^{-1}); when the prescribed maximum displacement was reached, the tip was held at peak load for 10 s and then unloaded. The load – displacement data obtained from the uniaxial compression tests were converted to engineering stress and strain. Since the deformation of sub–micron gold pillars is expected to be inhomogeneous [3], the engineering stress and strain can be used here. Because the diameter of the flat punch tip is considerably greater than the diameter of the nanowires, a large number of individual nanowires are deformed in parallel, in simple compression, during a single test. The average engineering stress in a single nanowire was calculated by the following equation:

$$\bar{\sigma} = \frac{4F}{N\pi d^2}, \quad (1)$$

where F is the load on the indenter, d is the mean diameter of the gold nanowires in the forest and N is the number of deformed nanowires. The number of deformed nanowires in each indent was determined from individual SEM images and the apparent "missing" wires visible in figure 2 accounted for.

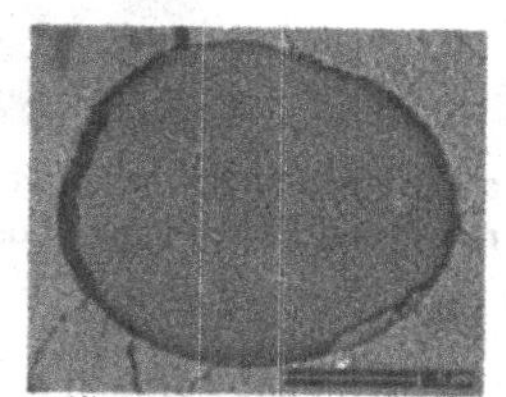

Figure 2. Residual indent obtained after testing a section of the nanoforest in compression.

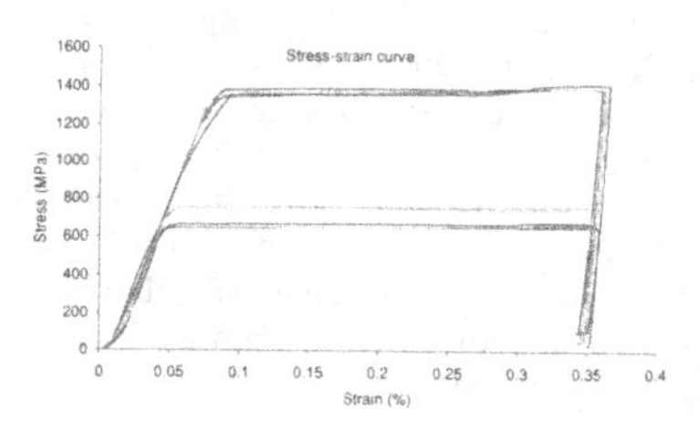

Figure 3. Engineering stress-strain curves from three gold nanoforests. The data shows three measurements taken for each nanowire diameter.

The engineering stress-strain curves of three nanowire forests with mean diameters 30, 60 and 70 nm are plotted in figure 3. The yield stress is taken as the plateau stress after 10% strain. No significant strain hardening is observed. The mean yield strength increases with decreasing nanowire diameter and the smallest gold nanowires, with 30 nm diameter, have a yield strength of 1.4 GPa.

TEM investigation

TEM investigation was conducted to observe the effect of plastic deformation on the nanowires. The deformed nanowires were released into 3 M NaOH solution and dispersed on a carbon coated TEM grid for observation. TEM image of a deformed nanowire is shown in figure 4a. The stepped surface was observed at the surface of the deformed nanowire which confirmed the plastic deformation by dislocation slip. Furthermore, deformation twinning was found in the deformed nanowire as shown in figure 4b. Therefore, the deformation is likely to be by conventional dislocation motion and possibly twinning.

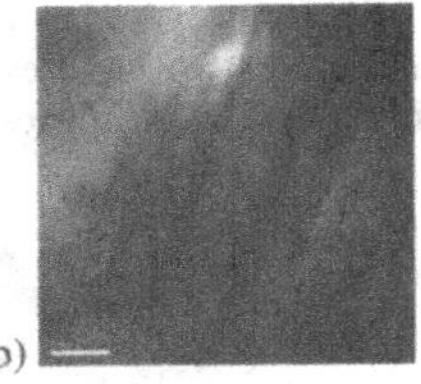

Figure 4. a) TEM image obtained from a deformed nanowire, inset diffraction pattern suggests possible deformation twinning. b) High resolution TEM image of deformation twinning in deformed nanowire.

Uniaxial compression test on nanoporous gold nanowores

We also fabricated nanoporous gold nanowires by using the same AAO film. Instead of electrodeposition of Au, Au-Ag alloys were deposited into the pores and then the Ag component was removed by selective dissolution in nitric acid. As shown in figure 5a, nanoporous gold nanowires with 5nm diameter ligaments and 20% relative density can be fabricated. The engineering stress-strain curves of nanoporous gold nanowires, obtained from identical uniaxial compression test on gold nanowires, are converted into engineering stress-strain curves of ligaments by using the model for open-cell foam mechanical behaviour developed by Gibson and Ashby [8]. (see figure 5b) The strain hardening due to strain gradients or densification can be observed and the yield point can be identified around 3 GPa.

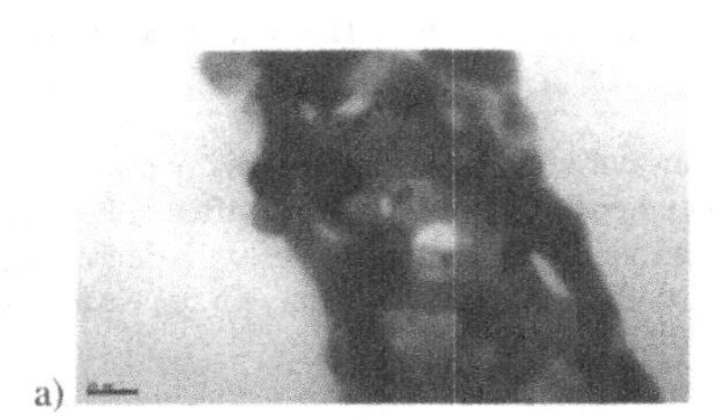

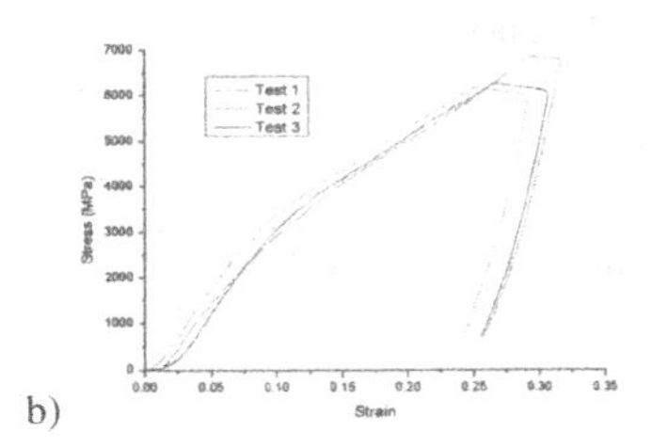

Figure 5. a) TEM image of nanoporous gold nanowire. b) Engineering stress-strain curves of ligament within nanoporous gold nanowires.

DISCUSSION

Discontinuous strain bursts in the stress – strain curves of single sub-micron gold pillars have been reported by several researchers [1-3]. But in our experiment, the flat punch tip is considerably larger than the nanowires so that the measurements represent the average values obtained from many nanowires compressed in parallel ($N > 10^4$). Thus the engineering stress – strain data show relatively less features than that obtained from single gold pillar. All of our gold nanowires show higher compression yield stresses than the yield stress of bulk gold with the smallest 30 nm-diameter gold nanowires having a high yield strength of 1.4 GPa. The yield strength data for the gold nanowire forests and previous results from the literature for the compression strength of gold pillars are plotted in figure 6. All of the experimental results show a strong size effect with the yield strength varying inversely with the diameter of the nanowires. The gold nanowire forest data lie on the same trend line obtained from the deformation of FIB machined gold pillars with a dependence of yield strength, σ_Y, on pillar or wire diameter, d, given by the following empirical relation:

$$\sigma_Y = Kd^{-0.6} \quad (2)$$

where K is a constant.

The strength of the ligaments within nanoporous gold has been deduced from its bulk mechanical properties using the model for open-cell foam mechanical behaviour developed by Gibson and Ashby [8]. By using this model, the calculated yield strength of the ligaments within nanoporous gold nanowire is still consistent with the same trend line obtained from the deformation of gold nanowires and FIB machined gold pillars (see figure 6).

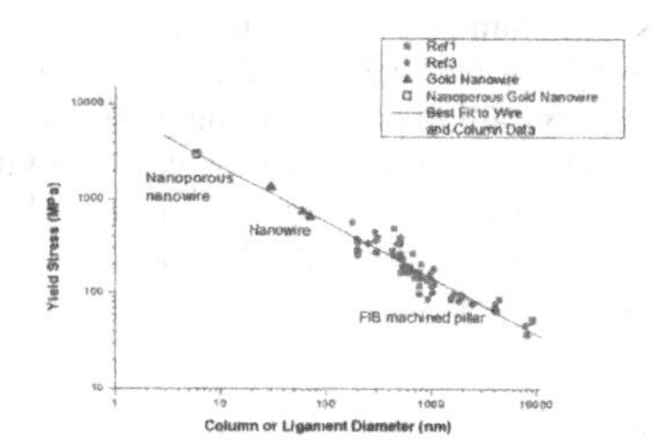

Figure 6. a) Engineering stress-strain curves from nanoporous gold nanowires. b) Yield stress as a function of diameter for the gold nanowires tested here in compression (red filled symbols), compression tests on larger columns produced by FIB milling (grey filled symbols), compression tests on nanoporous gold nanowore (red square symbol); Line shows linear fit to nanowire and nanocolumn data.

From figure 4a we can clearly identify the surface steps which are believed to be the intersection of active dislocation slip planes with the nanowire surface. No dislocations were observed in the deformed nanowires, and no significant hardening of the wires occurred. This can be explained by the dislocation starvation mechanism proposed by Nix et al [1-4]. In this mechanism the presence of the free surface of a nanowire generates image stresses that attract dislocations to the surface, where they leave behind a slip step, accounting for the low dislocation density in the material after compression testing. However, deformation twinning was found in high resolution TEM image (see figure 4b). Twinning has been predicted in molecular dynamic studies of gold nanowire deformation [9] and has been found in other studies of the deformation of small volumes of gold [10]. On the other hand, significant strain hardening was found in the engineering stress-strain curves of nanoporous gold nanowires. (see figure 5b) It has been known for some time that any gradient in plastic strain must be accommodated by a population of geometrically necessary dislocations [11] and that these dislocations will result in enhanced strengthening of the material or strain gradient hardening. Therefore, the observed strain hardening could be attributed to the presence of strain gradients during the deformation. However, further TEM investigations on deformed nanoporous gold nanowires are needed to confirm the deformation mechanism which operates in nanoporous gold nanowires.

There are two different mechanisms for the size dependent strength observed in nanoporous gold (strain gradients) and that observed during the compression of nanowires (dislocation starvation), and there is no evidence for these two structures having a common deformation mechanism. Thus, the correlation between the yield strength of ligament in nanoporous gold nanowires and the yield strength of gold nanowires and pillars clearly needs further investigation.

CONCLUSIONS

This study shows the compression yield strength of gold nanowires with diameters in the range 30 – 70 nm. The same dependence of the yield strength on nanowire diameter has been found as reported for gold pillars with diameter in the range 200 – 1000 nm. The maximum yield strength measured at approximately 1.4 GPa for 30 nm-diameter nanowires is significantly higher than that expected from polycrystalline gold specimens. However, deformation appears to be by conventional mechanisms of dislocation motion and possibly twinning. A strong correlation between the yield strength of ligament in nanoporous gold nanowires and the yield strength of gold nanowires and pillars has been observed, although the two structures are thought to have two different deformation mechanisms.

REFERENCES

1. J.R. Greer, W.C. Oliver and W.D. Nix, Acta Mater. 53 1821 (2005).
2. J.R. Greer and W.D. Nix, Phys. Rev. B. 73 245410 (2006).
3. C.A. Volkert and E.T. Lilleodden, Philos. Mag. 86 5567 (2006).
4. W.D. Nix, J.R. Greer, G.Feng and E.T. Lilleodden, Thin Solid Films 515 3152 (2007).
5. H. Masuda and K. Fukuda, Science 268 1466 (1995).
6. H. Masuda, H. Yamada, M, H. Asoh, M. Nakao and T. Tamamura, Appl. Phys. Lett. 71 2770 (1997).
7. R, Dou, and B. Derby. Scripta Mater. 59 151-154 (2008).
8. L.J. Gibson, M.F. Ashby. Proc. Royal Soc. Lon. A; 382A: 43-59 (1982).
9. H.S. Park, K. Gall and J.A.Zimmerman, J. Mech. Phys. Sol. 54 1862 (2006).
10. T. Kizuka, Phys. Rev. B. 57 11158 (1998).
11. M.F. Ashby. Philos. Mag. 21:399-424. (1970).

Mater. Res. Soc. Symp. Proc. Vol. 1144 © 2009 Materials Research Society 1144-LL21-11

Electrofluidic Positioning of Biofunctionalized Nanowires

Thomas J. Morrow[1], Jaekyun Kim[2], Mingwei Li[2], Theresa S. Mayer[2], Christine D. Keating[1*]

Departments of Chemistry[1] and Electrical Engineering[2], The Pennsylvania State University, University Park, PA 16802.
*Author to whom correspondence should be addressed at keating@chem.psu.edu

ABSTRACT

We functionalized nanowires with three different probe peptide nucleic acid (PNA) sequences, and assembled the three populations onto a lithographically patterned chip. Electrofluidic assembly enabled positioning each set of nanowires to span a different pair of guiding electrodes. Fluorescence imaging was used to probe whether the PNA on the individual nanowires remained able to selectively bind complementary DNA targets following assembly and integration of the positioned nanowires onto the chip surface.

INTRODUCTION

Chip-based nanowire sensing methods are promising candidates for ultraportable bioanalysis due to the small size of individual wires, and the potential for ultrasensitive electrical, electrochemical, and mass-based detection of biological target molecules. detection[1- 3]. Such a device would provide critical information in emergency situations, and could aid in the early diagnosis of complex diseases such as cancers or respiratory infections. Consequently there has been intense interest in developing on-chip methods such as nanowire field effect transistors (FETs), micro/nano electromechanical systems (MEMS or NEMS), and electronic noses[4-6]. The incorporation of nanowires functionalizd with probes for a large number of different targets (multiplexing) –while retaining the activity and selectivity of the probe molecules– is an obstacle currently limiting the further development of these sensors. Although some nanowire positioning [7, 8] and probe molecule delivery methods [9-11] have been developed to overcome this obstacle, these methods lack the ability to place nanowires in predetermined locations needed for highly multiplexed systems, have limited types of probe molecules that can be delivered, and cannot ensure optimal attachment chemistries of the probe molecules, possibly leading to cross contamination of adjacent sensing elements. For these reasons, we have developed a versatile electrofluidic positioning method which allows us to position pre-functionalized nanowires to predetermined chip locations, such as between columns of electrodes. Scheme 1 illustrates the concept.

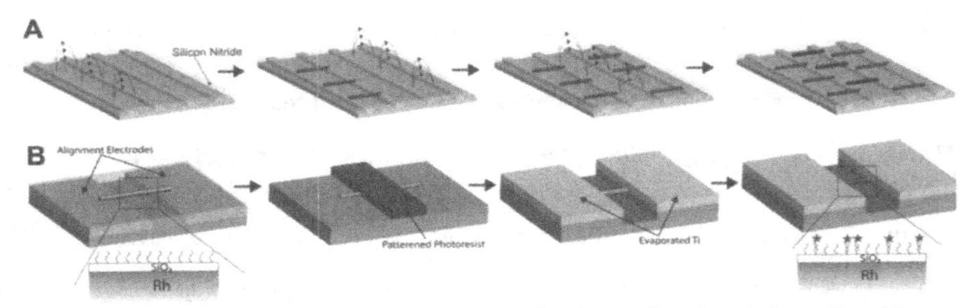

Scheme 1. (A) Selective positioning of pre-functionalized nanowires. Populations of peptide nucleic acid (PNA)/SiO_2 Rh nanowires were sequentially injected onto the chip surface while alternating voltages were applied to specific sets of guiding electrodes guiding the nanowires to the gap area. (B) Schematic illustrating the integration procedure of aligned nanowires. Following alignment a second photoresist was applied patterned exposing the ends of the nanowires while protecting the biomolecules functionalized to the middle segment. Ti was evaporated anchoring the exposed ends of the nanowires. Standard lift-off was used to remove excess Ti and to expose the biomolecules on the once protected middle segment of the nanowire to hybridize with their fluorescently labeled target DNA sequences.

EXPERIMENTAL DETAILS

Nanowire Synthesis

Metallic nanowires (8 μm in length) were synthesized by galvanostatic electrodeposition into the pores of aluminum oxide templates with a nominal pore diameter of 0.2 μm [12, 13]. The nanowires were rinsed by centrifugation and resuspended in 1 mL of ethanol resulting in $\sim 10^9$ nanowires. Rh nanowires were coated with ~30 nm of SiO_2 [14] and functionalized with PNA probe molecules (P1, P2 or P3 shown in Table 1) using previously described methods [5] and suspended in ethanol for alignment. We selected sequences that were complementary to nucleic acids from pathogenic viruses hepatitis C (HCV), hepatitis B (HBV), and human immunodeficiency virus (HIV). As targets, we used synthetic DNA oligonucleotides for these sequences, with different fluorescent tags on each target strand (Table 1).

Table 1

Name	Sequence 5'→3'	Description
P1	Thiol-TTTTTTTTTTGAGTAGTGTTGGGTCGCGAA	HCV PNA Probe
P2	Thiol-TTTTTTTTTTCTCAATCTCGGGAATCTCAA	HBV PNA Probe
P3	Thiol-TTTTTTTTTTCCATCAATGAGGAAGCTGCA	HIV PNA Probe
T1	[a]Alexa Fluor 488- TTCGCGACCCAACACTACTC	HCV DNA Target
T2	[b]TAMRA-TTGAGATTCCCGAGATTGAG	HBV DNA Target
T3	[c]Alexa Fluor 647-TGCAGCTTCCTCATTGATGG	HIV DNA Target

(a) λ_{ex}=482 nm, λ_{em}=517 nm; (b) λ_{ex}=559 nm, λ_{em}=583nm; (c) λ_{ex}=650 nm, λ_{em}=670 nm

Patterning of Silicon Wafers and Electrofluidic Assembly

Electrodes (32 μm wide, 5 mm long and separated by a 3 μm gap) were fabricated by standard metal liftoff of 10 nm of Ti, and 90 nm of Au on a 300 nm layer of thermally grown SiO_2 on a silicon (100) substrate. Standard lithographic techniques using PMGI-SF6 and Shipley 1811 in a dual layer resist process were used to pattern them. Silicon nitride was deposited to insulate the guiding electrodes during the assembly process.

Electrofluidic Assembly of the PNA Functionalized Nanowires Between Specific Electrodes

Alignment of the three sets of PNA functionalized nanowires was accomplished by applying a non-uniform electric field with a frequency of 1 MHz and a voltage of 3 V_{rms}. Nanowires functionalized with P1, P2 or P3 were deposited on the substrate and positioned into the desired gaps, allowing the ethanol to evaporate after each set of functionalized nanowires were positioned into the gap region.

PNA Hybridization Procedure

Hybridization of T1, T2 and T3 to their complementary probe molecules was accomplished by incubating the targets at a concentration of 0.38 μM in PBS buffer at room temperature for ~15 hours. The wafer was rinsed in excess PBS and a cover slip was added for imaging using an oil immersion 60x objective (1.4 NA). The fluorophores synthesized onto the target DNA sequences were chosen to minimize the spectral overlap of the absorbing and emitting wavelengths (listed under table 1).

DISCUSSION/RESULTS

Modeling the Positioning of Nanowires

We simulated the forces acting on the nanowires during assembly. Electric field calculations were conducted in which alternating voltages of 3V_{rms} at a frequency of 1MHz were applied between guiding electrodes (Figure 1). The area of highest electric field was at the edges of the guiding electrodes in the gap region. Dielectrophoretic forces drive the nanowires between the desired gaps in under 1 min. After a wire is positioned into a specific location, the driving force for additional nanowires to assembly nearby is greatly reduced, which results in a spacing of several microns between individual wires [5, 15].

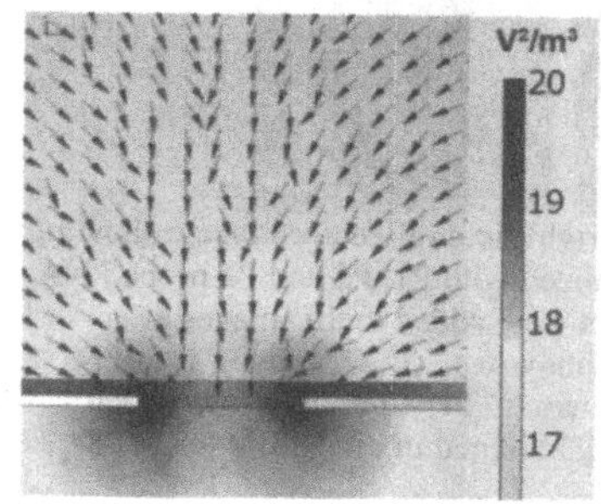

Figure 1. Cross-sectional view of a single gap (3 μm) plotted as $\log_{10} \nabla E^2$; the arrows show the direction of the long range dielectrophoretic forces.

Positioning of PNA Functionalized Nanowires

Three batches of SiO_2 coated Rh nanowires were functionalized with three different probe PNA sequences (P1, P2 and P3) in suspension prior to positioning them onto the chip. They were then positioned one batch at a time into pre-determined locations by applying non-uniform electric field gradients between desired electrodes. The ethanol suspending the populations of injected nanowires was allowed to fully evaporate before subsequent injections of the other nanowire populations. This helps ensure that when the electric field is removed and applied to other sets of guiding electrodes for subsequent injections, the nanowires positioned in prior steps remain in their desired locations rather than migrating to the newly activated gaps. Following sequential alignment of the three PNA probe coated Rh nanowire populations, a mixture of all three fluorescently labeled DNA targets (T1, T2 and T3) was incubated with the aligned nanowires to check whether probe molecules remained active and selective following the electrofluidic assembly process (electric fields, being suspended in ethanol). Figure 2 shows reflectance and fluorescence microscopy images of the three-column nanowire arrays after target hybridization. All of the nanowires can be seen in the reflectance image (far left). Wires from only one column are visible in each of the fluorescence images, which from left to right show the position of Alexa488-labeled T1, TAMRA-labeled T2, and Alexa647–labeled T3. Since a color overlay was not possible for this Proceeding, we have numbered the wires identically in each image to facilitate comparisons between them. These data show both that the fluorescently labeled target molecules hybridized selectively to their complementary PNA probe molecules on the wires (indicating retention of PNA function after assembly), and that the sequential nanowire assembly process led to the intended placement of different nanowire populations across the different electrode pairs.

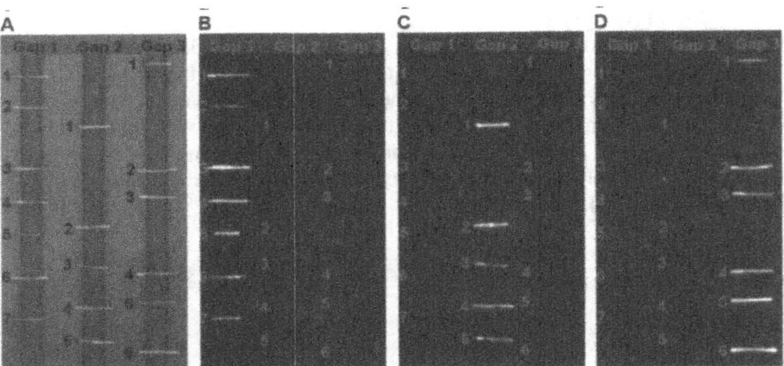

Figure 2. Positioning of PNA coated Rh nanowires. (A) Bright field reflectance image showing the positions of all three populations of nanowires functionalized with P1, P2 and P3 probe PNA aligned in Gap 1, Gap 2 and Gap 3 respectively, the numbers were added to the images after acquisition to easily determine the locations of the aligned nanowires. (B, C, and D) Fluorescence images showing that the fluorescently labeled targets T1, T2 and T3 hybridized selectively to the nanowires functionalized with P1, P2 and P3 aligned in image B(Alexa 488), image C (TAMRA) and image D (Alexa 647), respectively.

Integration of Biofunctionalized Nanowires

An important challenge for the bottom-up assembly of nanoscaled building blocks is integration with the on-chip circuitry to form arrays with a high density of devices. To address this problem, biocompatible parameters were developed for use with standard photolithographic techniques to maintain the activity and selectivity of the sensitive PNA molecules functionalized to the nanowires. Following alignment as described above, photoresist was applied over top of the substrate and the aligned nanowires. The photoresist helps protect the PNA probe molecules bound to the central section of each nanowires while electrical contact is made to the nanowire ends. This is done by patterning windows over the ends of the aligned nanowires, into which Ti is evaporated, anchoring the aligned nanowires. Following evaporation, standard lift-off procedures were used, exposing the PNA probe molecules to their environment. To ensure that the PNA remains active and selective following nanowire anchoring, fluorescently labeled targets T1, T2 and T3 were again incubated in PBS buffer to hybridize to their complementary PNA probes functionalized to the aligned nanowires. Figure 3 shows the results of this experiment. As in Figure 2, where no post-assembly processing was performed, the fluorescently labeled targets T1, T2 and T3 successfully hybridized to their complementary PNA probe molecules showing that the PNA remains both active and selective following both the electrofluidic alignment and integration procedures. Some of what appear to be wires in the reflectivity image shown in panel A are in fact regions where nanowires have been lost during processing, leaving behind Ti-free "shadows"; these have not been numbered, since they cannot be expected to show up in any of the fluorescence images. Improved clamping procedures should prevent this in the future. Importantly, these data show that standard photolithographic techniques can be compatible with biorecognition probe molecules such as PNA, paving the way for biocompatible photolithographic methods that will be required to increase the (bio)functionality of integrated circuits.

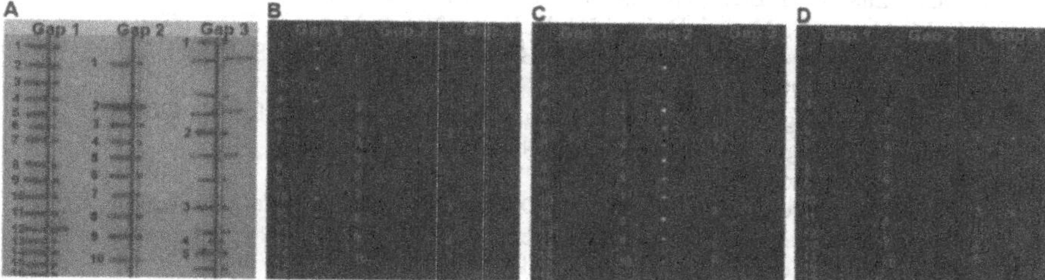

Figure 3. (A) Bright field reflectance image showing the positions of all three populations of nanowires functionalized with P1, P2 and P3 probe PNA aligned in Gap 1, Gap 2 and Gap 3, respectively following integration with evaporated Ti (~300nm). (B, C, and D) Fluorescence images showing that the fluorescently labeled targets T1, T3 and T2 hybridized selectively to the nanowires following integration procedures with P1, P2 and P3 aligned in Gap 1 (Alexa 488), Gap 2 (TAMRA) and Gap 3 (Alexa 647) respectively.

CONCLUSIONS

Bottom-up assembly is a promising route to for the incorporation of different materials or molecules with lithographically defined features. The electrofluidic assembly method described here shows that it is possible to position nanowire building blocks to pre-determined locations on the chip surface. By using pre-functionalized nanowires and positioning them to predetermined locations on the chip surface, we hope to increase the multiplexing capabilities of chip-based sensors, which will have a tremendous impact in emergency situations, and increase the functionality of integrated chip devices revolutionizing the chemical and biomedical industries.

ACKNOWLEDGMENTS

This work was supported by the National Institutes of Health (R03 CA118591 and R01 EB00268) and the National Science Foundation (CCR-0303976). The authors also acknowledge use of facilities at the PSU site of NSF NNIN.

REFERENCES

1. F. Patolsky, G. Zheng, C. M. Lieber, Anal. Chem., 78, 4260 (2006).
2. J. Wang, *Analyst,* **130,** 421, (2005).
3. L. G. Carrascosa, M. Moreno, M. Álvarez, L. M. Lechuga, Trends in Anal. Chem., 25, 196 (2006).
4. F. Patolsky, B. P. Timko, G. Zheng, C. M. Lieber, MRS Bull., 32, 142 (2007).
5. M. Li, R. B. Bhiladvala, T. J. Morrow, J. A. Sioss, K. Lew, J. M. Redwing, C. D. Keating, T. S. Mayer, Nature Nanotech., 3, 88 (2008).
6. Y. Cao, A. E. Kovalev, R. Xiao, J. Kim, T. S. Mayer, T. E. Mallouk, Nano Lett. In Press 10.1021/nl800940e (2008).
7. G. Yu, A. Cao, C. M. Lieber, Nature Nanotech., 2, 372 (2007).
8. M. C. McAlpine, H. Ahmad, D. Wang, J. R. Heath, Nature Mater. 6, 379 (2007).
9. R. P. Aubrun, D. P. Kreil, L. A. Meadows, B. Fischer, S. S. Matilla, S. Russel, Trends Biotech., 23, 374 (2005).
10. A. Bietsch, J. Zhang, M. Hegner, H. P. Lang, C. Gerber, Nanotech., 15, 873 (2004).
11. K. Salaita, Y. Wang, C. A. Mirkin, Nature Nanotech.,2, 145 (2007).
12. C. R. Martin, Science, 266, 1961 (1994).
13. S. R. Nicewarner-Pena, R. G. Freeman, B. D. Reiss, L. He, D. J. Pena, I. D. Walton, R. Cromer, C. D. Keating, M. J. Natan, Science, 294, 137 (2001).
14. J. A. Sioss, C. D. Keating, Nano Letters, 5, 1779 (2005).
15. P. A. Smith, C. D. Nordquist, T. N. Jackson, T. S. Mayer, B. R. Martin, J. Mbindyo, T. E. Mallouk, Appl. Phys. Lett. 77, 1399 (2000).

AUTHOR INDEX

SUBJECT INDEX

Printed in the United States
By Bookmasters